Physics and Chemistry in Space Vol.11

Edited by L. J. Lanzerotti, Murray Hill

Hans Volland

Atmospheric Electrodynamics

With 120 Figures

Springer-Verlag
Berlin Heidelberg New York Tokyo 1984

Professor Hans Volland

Radioastronomisches Institut
Universität Bonn
Auf dem Hügel 71
5300 Bonn, FRG

ISBN-13: 978-3-642-69815-6   e-ISBN-13: 978-3-642-69813-2
DOI: 10.1007/978-3-642-69813-2

Library of Congress Cataloging in Publication Data. Volland, Hans. Atmospheric electrodynamics. (Physics and chemistry in space ; v. 11) Bibliography: p. 1. Atmospheric electricity. 2. Electrodynamics. 3. Magnetosphere. 4. Space plasmas. I. Title. II. Series.
QC801.P46 vol. 11 [QC961] 523.01 s [551.5′6] 84-10582

© by Springer-Verlag Berlin Heidelberg 1984
Softcover reprint of the hardcover 1st edition 1984

Typesetting: Fotosatz GmbH, Beerfelden. Offsetprinting and Bookbinding: Konrad Triltsch,
Graphischer Betrieb, Würzburg.
2131/3130-543210

Preface

This book resulted from lectures which I gave at the Universities of Kyoto, Cologne, and Bonn. Its objective is to summarize in a unifying way two otherwise rather separately treated subjects of atmospheric electrodynamics: electric fields of atmospheric origin, in particular thunderstorm phenomena and related problems on the one hand, and magnetic fields, in particular those which are associated with electric currents of upper atmospheric origin, on the other. Geoelectricity and geomagnetism were not always considered as belonging to quite different fields of geophysics. On the contrary, they were recognized by the physicists of the 19th and the beginning of the 20th century as two manifestations of one and the same physical phenomenon, which we presently refer to as electromagnetic fields. This can still be visualized from the choice of names of scientific journals. For instance, there still exists the Japanese *Journal of Geomagnetism and Geoelectricity,* and the former name of the present American Journal of Geophysical Research was *Terrestrial Magnetism and Atmospheric Electricity.*

Whereas geomagnetism became the root of modern magnetospheric physics culminating in the space age exploration of the earth's environment, geoelectricity evolved as a step-child of meteorology. The reason for this is clear. The atmospheric electric field observed on the ground reflects merely the local weather with all its frustrating unpredictability. The variable part of the geomagnetic field, however, is a useful indicator of ionospheric and magnetospheric electric current systems.

Only in the last two decades have ionospheric and magnetospheric physicists rediscovered the importance of electric fields of upper atmospheric origin. With the development of new instruments and their carriers (balloons, rockets, satellites), electric and magnetic fields of lower and upper atmospheric origin are now measured within the whole atmosphere from the ground to the magnetosphere and beyond. This again closes the gap between geoelectricity and geomagnetism that existed for more than a half a century. It is appropriate to select a new name for this branch of research: Atmospheric Electrodynamics.

In order to deal with such a broad field within the limiting space of a monograph, severe restrictions were necessarily imposed on the selection of the topics. Naturally, this selection is somewhat biased toward the author's own work. The potential reader is expected to possess some basic knowledge in electrodynamics. In the quantitative treatments of the subjects, simple analytical solutions are generally preferred in order to provide the reader with a physical insight into the problem rather than a full sophisticated description of the phenomenon in detail.

I wish to express my thanks for helpful advice and suggestions to S. I. Akasofu, W. Baumjohann, G. Becker, M. K. Bird, A. Egeland, S. Grzedzielski, P. Ingmann, M. Kreitz, W. Koehnlein, L. J. Lanzerotti, M. C. Maynard, G. W. Prölss, A. D. Richmond, G. Rostoker, J. Schäfer, M. Schmolders, and S. D. Shawhan.

Bonn, Juli 1984   *Hans Volland*

Contents

1 Introduction

Benjamin Franklin's famous experiment in 1752 "drawing lightning from the cloud" by a kite (e.g., Dibner 1977) is generally considered as the beginning of the science of Atmospheric Electricity. The term "Atmospheric Electricity" reflects the earlier efforts to study mainly the electrostatic component of the geo-electromagnetic field. This subject now includes thunderstorm electrification, the global electric circuit, but also lightning and related phenomena, and the propagation of lightning pulses in the atmosphere (sferics).

Prior to the space age, the altitude accessible for in situ experiments was limited by the height which balloons could reach (≈ 30 km). Since the observations showed a nearly exponential increase with altitude of the electric conductivity of the air, it was reasonable to speculate about an electric equipotential layer – the "electrosphere", or in German, "Ausgleichsschicht" – assumed to exist above 40 to 60 km. This electric equipotential layer acts as the upper plate of a huge spherical condenser, the earth's surface being the lower plate. The region between these two boundaries is the realm of conventional atmospheric electricity (Chalmers 1967; Israel 1970, 1973; Dolezalek and Reiter 1977).

Whistlers, however, which are electromagnetic pulses generated by lightning events that propagate along the geomagnetic field lines into the magnetosphere, thereby penetrating the imaginary electrosphere, clearly ignore this artificial upper boundary layer. Moreover, electric fields of magnetospheric origin have been measured at balloon altitudes, well below that imaginary boundary. This leads to a broader view which now includes low frequency electric and magnetic fields generated within the lower and upper atmosphere. These electromagnetic fields can penetrate the entire atmosphere, reaching from the surface of the earth to the highest fringe of the atmosphere – the magnetosphere.

Since the air is electrically conducting, an electric field in the atmosphere cannot be maintained indefinitely, but must be generated by nonelectric forces. Three sources of low frequency electromagnetic waves are presently known:

1. Thunderstorms and related phenomena in the lower atmosphere.
2. Tidal wind interaction with the ionospheric plasma at dynamo layer heights.
3. Solar wind interaction with the magnetosphere.

While source 1 is the main subject of conventional atmospheric electricity, source 2 has long been studied by geomagnetic and ionospheric people dealing with the so-called Sq and L currents at ionospheric heights and their magnetic signature at the earth's surface (Chapman and Bartels 1951; Matsushita and Campbell 1967; Kato 1980). Source 3, on the other hand, involves the topic of magnetospheric physics, the knowledge of which has dramatically increased during the recent decades of space exploration (Akasofu and Chapman 1972; Akasofu 1977; Kennel et al. 1979). This source generates large-scale electric fields which are related to global-scale plasma convection. These fields map down into the ionosphere and drive electric currents within the dynamo region, the magnetic component of which can be observed on the ground. In addition, instabilities within the magnetosphere are the cause of electromagnetic waves with frequencies ranging from the 0.1 mHz band to the kHz band and beyond (geomagnetic substorms, geomagnetic pulsations, and natural ELF, VLF and LF noise).

This book describes these three sources and their low frequency electromagnetic radiation in the atmosphere from a common point of view. The limitation to low frequency waves is necessary to keep the subject within reasonable bounds. In general, only frequencies below about 100 kHz will be treated. This is the range where individual wave structure of quasi-harmonic or impulsive nature can still be separated. Signals beyond that frequency resemble mostly a continuous noise spectrum. The term "Atmospheric Elec-

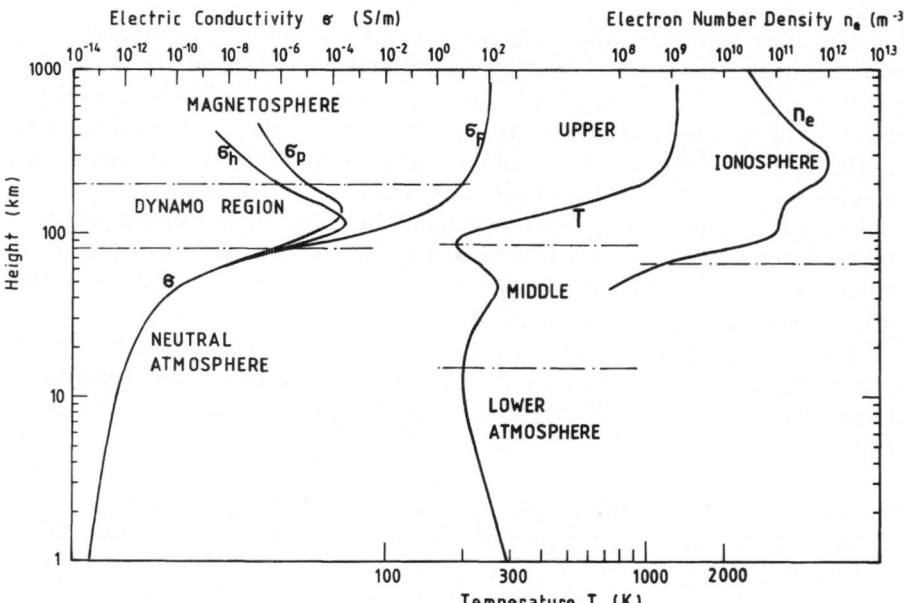

Fig. 1.1. Nomenclature of atmospheric regions based on profiles of electric conductivity, neutral temperature, and electron number density

trodynamics" was coined in order to emphasize the broader scope of the topics as compared with conventional atmospheric electricity.

Throughout this book, several nomenclatures for the various atmospheric levels will be used related to different atmospheric parameters. Altitude regions selected according to their neutral temperature profile are the lower atmosphere (troposphere: <15 km), middle atmosphere (stratosphere and mesosphere: 15 to 85 km), and upper atmosphere (thermosphere and exosphere: >85 km) (see Fig. 1.1). The same height range can be divided according to its plasma state into the neutral atmosphere (<80 km), dynamo region (80 to 200 km), and magnetosphere (>200 km). The regime of enhanced electron density above about 65 km height is called the ionosphere. The region of maximum electron density near 250 km height is the F region. The relative maximum near 100 km altitude is the E region of the ionosphere.

2 Plasma Component of the Air

The atmosphere is an electrically conducting medium so that electric currents can flow. Since the electric conductivity σ is a basic parameter for electromagnetic fields and currents, we outline in this chapter the origin and distribution of σ within the atmosphere.

2.1 Sources of Ionization

The earth's crust contains radioactive material, mainly uranium, thorium, and their decay products. Beta and gamma rays emitted from the ground can ionize the molecules of the air in the first few meters above ground. The gas radon, which is one of the decay products of uranium 238 can reach greater heights, up to several 100 m above ground, before it decays into polonium by emitting alpha particles. Radon is therefore a major ionization source in the

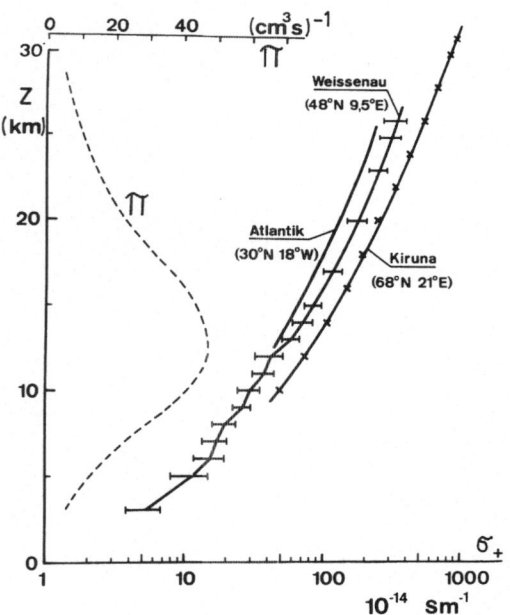

Fig. 2.1. Observed polar electric conductivity σ_+ due to the small positive ions at three different latitudes from balloon measurements, and ion production rate Π versus height. (Gringel 1977)

first few 100 m above ground over the continents. The ionization production rate on the ground is of the order of $\Pi \simeq 10^7 \, \text{m}^{-3} \, \text{s}^{-1}$.

Radioactivity in the oceans is several orders of magnitude smaller than in the continental crust and is thus of minor importance as an ionization source there (Israel 1970).

The second major source of ionization is galactic cosmic rays. The ionization production rate due to cosmic rays has a maximum of about $\Pi \simeq 4 - 5 \times 10^7 \, \text{m}^{-3} \, \text{s}^{-1}$ at midlatitudes near 15 km height (Fig. 2.1), decreasing to about $\Pi \simeq 10^6 \, \text{m}^{-3} \, \text{s}^{-1}$ at sea level. These rates decrease by a factor of about two at the equator. A solar cycle variation of about 20% with maximum ionization rates during sunspot minimum has been observed. The ionization rate is sensitive to meteorological conditions, and also to solar activity (Herman and Goldberg 1978).

Solar X-ray and extreme ultraviolet (XUV) radiation are the principle sources of ionization above 60 to 70 km altitude. Their ionization rates depend on latitude, time of day and season, and solar activity. Other sources of ionization in this height region are irregularly occurring highly energetic electrons and protons of magnetospheric and solar origin (Whitten and Poppoff 1971). During extremely intense solar flare events, solar protons of cosmic ray intensity can greatly enhance the ion production rate down to heights of 15 km.

2.2 Ion Composition

The ionization sources separate the molecules of the neutral air into ions and electrons. The ions are primarily singly charged. In the lower and middle atmosphere, the electrons are attached to neutral molecules forming negative ions. Positive and negative ions form molecular clusters via a hydration process within a few microseconds. Examples of such clusters, called small or fast ions, are $H_3O^+(H_2O)_n$, $H^+(H_2O)_n$, $O_2^-(H_2O)_n$, and $CO_4^-(H_2O)_n$, with $n \simeq 4 - 8$, depending on the water vapor content of the air (Mohnen 1977; Arnold 1980). The small ions disappear either by recombination with each other or by attachment to aerosol particles, thus forming long-living, relatively immobile, large ions.

The number of the small ions can be derived from an equation of recombination

$$dn/dt = \Pi - \alpha n^2 - \beta n_a n, \tag{2.1}$$

with n the number density of the small ions, Π the ion production rate, α a recombination coefficient, β an attachment coefficient, and n_a the number density of the aerosol particles. Equation (2.1) preassumes that the number densities of the positive and negative ions are nearly equal, which is normally a good approximation at all heights.

During steady state conditions ($dn/dt = 0$), n can be determined as a function of Π and n_a:

$$n \simeq [-\beta n_a + (\beta^2 n_a^2 + 4\alpha\Pi)^{1/2}]/(2\alpha) . \tag{2.2}$$

The values of α and β in the troposphere are of the order of

$$\alpha \simeq 2 \times 10^{-12} \, m^3 \, s^{-1}, \quad \beta \simeq 1 \times 10^{-11} \, m^3 \, s^{-1} .$$

The number density of the aerosol particles is highly variable, depending on orography, meteorological conditions, and man-made air pollution. Typical values on the ground are $n_a \simeq 10^{10} \, m^{-3}$ over the continents, and $10^9 \, m^{-3}$ over the oceans. The average number density of the small ions is roughly the same over the continents as over the oceans ($n \simeq 10^8 - 10^9 \, m^{-3}$), because the smaller ionization rate over the oceans is compensated by the smaller loss rate due to the lower aerosol concentration there (Israel 1970).

Water vapor becomes of minor importance at heights above 70 to 80 km. The ion clusters are therefore replaced mainly by singly charged positive ions like O^+, NO^+, O_2^+, and in greater heights H^+, and by the (negatively charged) electrons. The solar XUV radiation ($\lambda < 0.12 \, \mu m$) is responsible for the formation of various ionospheric layers with a maximum electron density of the order of $n \simeq 10^{12} \, m^{-3}$ within the ionospheric F2 layer near 250 km altitude (Fig. 1.1). The principle ion in that height range is O^+ (Thomas 1982).

2.3 Electric Conductivity of Lower and Middle Atmosphere

The positive and negative ions in the atmosphere are accelerated in opposite directions by an electric field. During their motion, they collide with the neutral particles. Their ability to move through the neutral gas is described by a mobility factor k which depends on the number density of the neutral air. The mobility of the ions increases with altitude in inverse proportion to the nearly exponential decrease of the air density. At ground level, the mobility factor of the positive small ions is about $k_+ \simeq 1.4 \times 10^{-4} \, m^2/(Vs)$ and somewhat larger for the negative small ions. The mobility of the large ions is several orders of magnitude smaller. Therefore, the large ions do not contribute significantly to charge transport.

The electric conductivity of the atmosphere below about 70 km height is isotropic. It depends on the product of the ion density and the mobility factor:

$$\sigma = \sigma_+ + \sigma_- = en(k_+ + k_-), \tag{2.3}$$

where $e = 1.6 \times 10^{-19} \, C$ is the elementary electric charge. Positive and negative small ions contribute nearly equally to the electric conductivity. Figure 2.1 shows the electric conductivity due to the positive small ions versus height at three different latitudes. Since the ion density varies only slightly with height, it is the increase of the ion mobility which is responsible for the nearly exponential increase of σ_+ with height. The increase of the conductivity with increasing latitude can be attributed to the increase of the ion density with latitude, which is related to the latitudinal dependence of the cosmic ray

Table 2.1. Coefficients of the parameters of the global electric circuit approximating average conditions at midlatitudes and at heights below about 60 km

i =	1	2	3	A
α_i (km^{-1})	4.527	0.375	0.121	
ϱ_i (10^{12} Ωm)	46.9	22.2	5.9	75.0
E_i (V/m)	93.8	44.4	11.8	150.0
Φ_i (kV)	20.7	118.4	97.5	236.6
Γ_i (10^{15} Ωm^2)	10.4	59.2	48.8	118.4
q_i (10^{-12}C/m^3)	3.76	0.147	0.013	3.92

intensity. During strong solar events, the conductivity in the middle atmosphere can increase by up to a factor of 10 (Reagan et al. 1983).

The average height dependence of the specific resistance of the atmosphere ϱ, which is defined as the reciprocal value of σ, can be analytically described by the formula

$$\varrho(z) = 1/\sigma(z) = \varrho_1 \exp(-\alpha_1 z) + \varrho_2 \exp(-\alpha_2 z) + \varrho_3 \exp(-\alpha_3 z) \qquad (2.4)$$

(z in km; ϱ in Ωm; ϱ_i and α_i in Table 2.1).

This formula is valid at midlatitudes below about 60 km altitude within the fair weather regions outside thunderstorms or cloudy areas.

From Eq. (2.4) follows a value of σ at sea level of

$$\sigma_A = 1/\varrho(0) = 1.33 \times 10^{-14} \, \text{S/m} . \qquad (2.5)$$

The observed electric conductivity can vary irregularly by more than a factor of 10, in particular within the atmospheric boundary layer in the first kilometer above ground, but also at stratospheric heights.

Integrating Eq. (2.4) over the height, one obtains a columnar resistance between sea level and a height z given by (see Fig. 2.2)

$$\Gamma = \int_0^z \varrho \, dz = \sum_{i=1}^{3} \Gamma_i [1 - \exp(-\alpha_i z)] , \qquad (2.6)$$

with $\Gamma_i = \varrho_i / \alpha_i$ (Table 2.1).

The columnar resistance between sea level and ionosphere ($z \simeq 100$ km) is then

$$\Gamma_A = \sum_{i=1}^{3} \Gamma_i = 1.18 \times 10^{17} \, \Omega \text{m}^2 . \qquad (2.7)$$

The columnar resistance is smaller over mountains.

The total resistance of the atmosphere becomes

$$R_A = \Gamma_A / F_E \simeq 230 \, \Omega , \qquad (2.8)$$

where $F_E = 5.1 \times 10^{14} \, \text{m}^2$ is the area of the earth's surface.

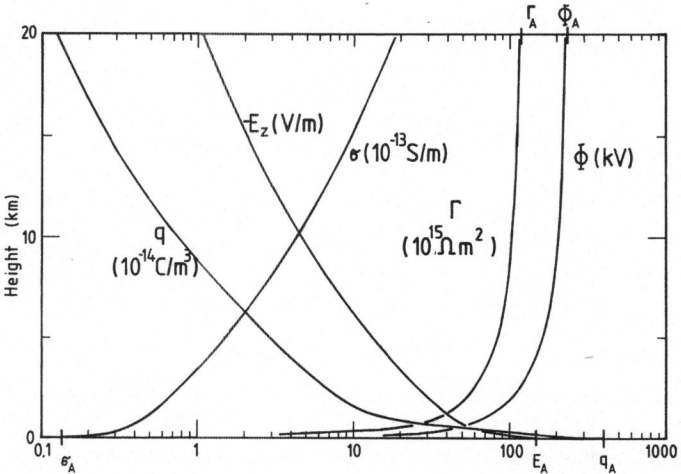

Fig. 2.2. Height profiles of the electric field E_z, electric potential Φ, electric conductivity σ, charge density q, and columnar resistance Γ (see Table 2.1)

2.4 Electric Conductivity of Upper Atmosphere

The electric conductivity in the ionosphere is anisotropic because the mobility of the ions and electrons depends on the direction of the geomagnetic field B_0. The mobility of the ions (electrons) parallel to the geomagnetic field is

$$k_\pm = e/(m_i v_i) \quad \langle e/(m_e v_e)\rangle, \tag{2.9}$$

where $m_i \langle m_e \rangle$ is the mass of the ions (electrons) and $v_i \langle v_e \rangle$ is the collision rate between one ion (electron) and the neutral particles. v_e is about two orders of

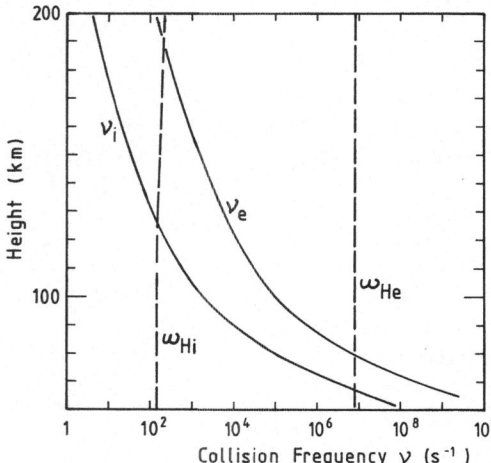

Fig. 2.3. Typical height profiles of collision frequencies v_i, v_e, and gyrofrequencies ω_{Hi}, ω_{He} of ions and electrons, respectively, at dynamo region heights

magnitude larger than v_i (Fig. 2.3), but m_i typically is more than four orders of magnitude larger than m_e. Thus,

$$m_i v_i \simeq 500\, m_e v_e, \tag{2.10}$$

so that the electrons dominate in the charge transport. The electric conductivity parallel to B_0, referred to as the parallel conductivity, may be written as

$$\sigma_F = \varepsilon_0 (\omega_i^2 / v_i + \omega_e^2 / v_e) \tag{2.11}$$

with $\omega_i = [e^2 n/(m_i \varepsilon_0)]^{1/2}$; $\langle \omega_e = [e^2 n/(m_e \varepsilon_0)]^{1/2} \rangle$ the plasma frequency of the ions ⟨electrons⟩, $\varepsilon_0 = 8.854 \times 10^{-12}$ F/m the permittivity of free space, and $n \simeq n_i \simeq n_e$ the number density of ions and electrons, respectively.

Ions and electrons move in circular orbits orthogonal to a homogeneous magnetic field B_0 if no collisions occur (Fig. 2.4). The gyrofrequency of the ions (electrons) is

$$\omega_{Hi} = e|B_0|/m_i; \quad \langle \omega_{He} = e|B_0|/m_e \rangle. \tag{2.12}$$

Therefore $\omega_{He} \simeq 5 \times 10^4\, \omega_{Hi}$ (Fig. 2.3).

If an electric field E is applied orthogonally to B_0, the ions and the electrons move orthogonally to E and B_0 with the same drift velocity [Fig. 2.4 and Eq. (14.20)]

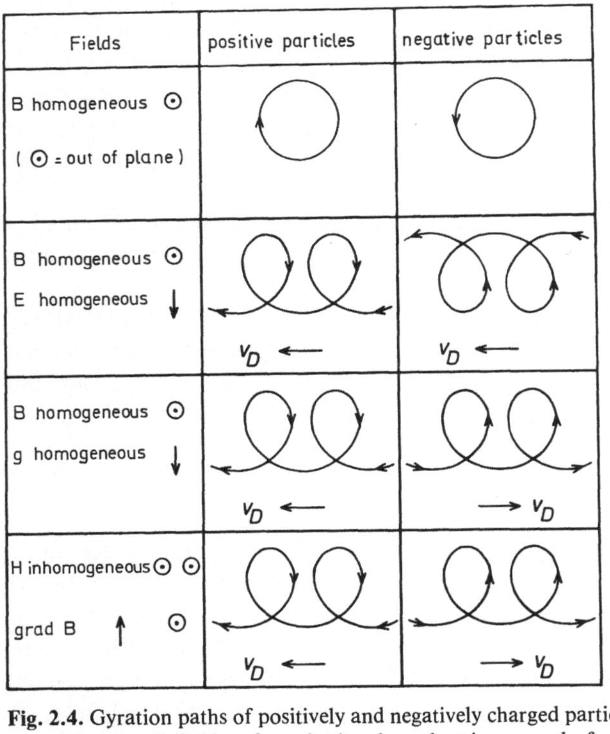

Fields	positive particles	negative particles
B homogeneous ⊙ (⊙ = out of plane)		
B homogeneous ⊙ E homogeneous ↓	v_D ←	v_D ←
B homogeneous ⊙ g homogeneous ↓	v_D ←	→ v_D
H inhomogeneous ⊙ ⊙ grad B ↑ ⊙	v_D ←	→ v_D

Fig. 2.4. Gyration paths of positively and negatively charged particles under the influence of electric and magnetic fields, of gravitational acceleration g, and of gradient of B. (Kertz 1969)

$$v_D = E \times B_0/B_0^2 . \tag{2.13}$$

If $v_i \gg \omega_{Hi}$, the movement of the ions is determined by the neutral gas, and the influence of the magnetic field is minor. On the other hand, if $v_i \ll \omega_{Hi}$, the influence of the geomagnetic field dominates. The electrons behave correspondingly. At an altitude of about 100 km, the situation $\omega_{He} \gg v_e$, but $\omega_{Hi} \ll v_i$ is attained (Fig. 2.3), so that the electrons drift orthogonally to B and E, while the ions still move in the direction of E.

The conductivity parallel to E but orthogonal to B_0, which is called the Pedersen conductivity σ_p, is mainly due to the ions:

$$\sigma_p = \varepsilon_0[\omega_i^2 v_i/(v_i^2 + \omega_{Hi}^2) + \omega_e^2 v_e/(v_e^2 + \omega_{He}^2)] . \tag{2.14}$$

The conductivity orthogonal to both E and B_0, mainly due to the electrons, is called the Hall conductivity σ_h:

$$\sigma_h = \varepsilon_0[-\omega_i^2 \omega_{Hi}/(v_i^2 + \omega_{Hi}^2) + \omega_e^2 \omega_{He}/(v_e^2 + \omega_{He}^2)] . \tag{2.15}$$

At heights above about 160 km, collisions with neutrals are much less frequent, so that $v_i \ll \omega_{Hi}$ and $v_e \ll \omega_{He}$. Thus,

$$\sigma_p \simeq nm_i v_i/B_0^2 \quad \text{and} \quad \sigma_h \simeq \varepsilon_0(\omega_e^2/\omega_{He} - \omega_i^2/\omega_{Hi}) = 0 . \tag{2.16}$$

The height range between 80 and 200 km, where σ_p and σ_h attain their peak amplitudes, is called the dynamo region (see Fig. 1.1). Since the parallel conductivity in the upper atmosphere is very large (much larger than the Pedersen and Hall conductivities), the geomagnetic field lines behave almost like electric equipotential lines, and electric fields parallel to B_0 break down within a fraction of a second. Significant electric currents can thus only flow if an electric field orthogonal to B_0 exists and if σ_p and σ_h are sufficiently large. This is the case in the dynamo region. The electric currents orthogonal to B_0 are of minor importance above that region. Likewise, the electric currents below 80 km altitude are small compared with the currents within the dynamo region.

The direction of the geomagnetic field changes from horizontal at the magnetic dip equator to vertical at the magnetic poles (see Sect. 14.2). A horizontal electric field near the poles can then drive horizontal Pedersen and Hall currents which are related to the electric field als

$$j = \sigma \cdot E \quad \text{with} \quad \sigma = \begin{pmatrix} \sigma_p & \sigma_h \\ -\sigma_h & \sigma_p \end{pmatrix} . \tag{2.17}$$

However, a west–east electric field E_y at the dip equator generates a Pedersen current in the y direction $[(j_p)_y = \sigma_p E_y]$ and a Hall current in the vertical (z) direction $[(j_h)_z = -\sigma_h E_y]$ (Fig. 2.5). Since the vertical current cannot flow out of the dynamo region, a polarization charge is built up on both boundaries of the dynamo region. This polarization charge causes a vertical electric polarization field $[E_z = (\sigma_h/\sigma_p)E_y]$ to compensate the Hall field, so that finally no vertical current flows $[j_z = 0]$. This secondary vertical field,

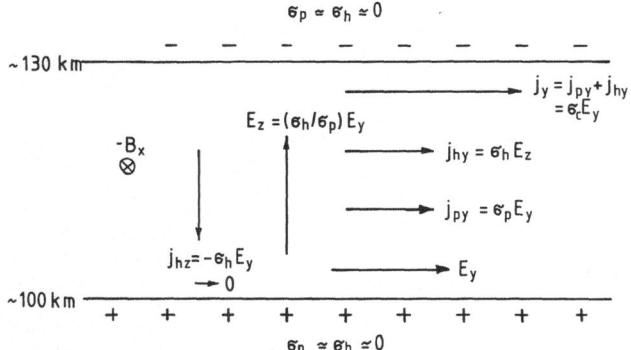

Fig. 2.5. Amplification of electric current at the geomagnetic equator within the dynamo region (Cowling current). (Coordinate system of Fig. 14.1c is adopted)

however, drives a horizontal Hall current $[(j_h)_y = \sigma_h E_z = (\sigma_h^2/\sigma_p) E_y]$ in the y direction which adds to the primary Pedersen current. This enhancement of the horizontal current can be described by the Cowling conductivity σ_c:

$$\sigma_c = \sigma_p + \sigma_h^2/\sigma_p. \tag{2.18}$$

The generalization of the condition $j_z = 0$ leads to a horizontal conductivity tensor in the dynamo region that depends on the geomagnetic dip angle I (Whitten and Poppoff 1971):

$$\sigma = \begin{pmatrix} \sigma_{xx} & \sigma_{xy} \\ -\sigma_{xy} & \sigma_{yy} \end{pmatrix} \tag{2.19}$$

with

$$\sigma_{xx} = \sigma_F \sigma_p/D$$
$$\sigma_{xy} = \sigma_F \sigma_h \sin I/D$$
$$\sigma_{yy} = [\sigma_F \sigma_p \sin^2 I + (\sigma_p^2 + \sigma_h^2)\cos^2 I]/D$$
$$D = \sigma_F \sin^2 I + \sigma_p \cos^2 I$$
$$\tan I = 2\tan\phi_m$$

where ϕ_m is the geomagnetic latitude for the case of a dipole configuration of the geomagnetic field (see Sect. 14.2), and the ionospheric coordinate system of Fig. 14.1c is valid in Eq. (2.19). It can be verified that $\sigma_{yy} \to \sigma_c$; $\sigma_{xy} \to 0$ for $I \to 0°$ (equator), and $\sigma_{xx} = \sigma_{yy} \to \sigma_p$; $\sigma_{xy} \to \pm\sigma_h$ for $I \to \pm 90°$ (poles). σ_h changes sign in the southern hemisphere because the dip angle becomes negative.

For many applications, the dynamo region can be approximated by a thin homogeneous layer of thickness $\Delta h \simeq 50$ km. The height-integrated conductivity of this layer is given by

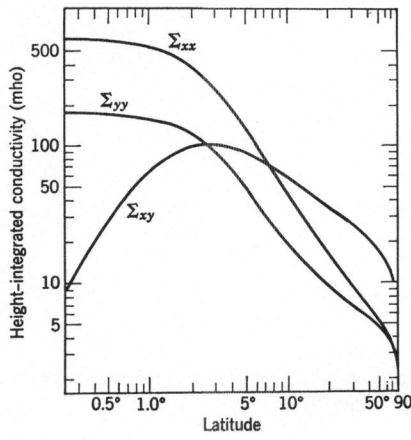

Fig. 2.6. Coefficients of the height integrated electric conductivity tensor Σ versus latitude (Fejer 1964). (Coordinate system of Fig. 14.1c is adopted)

$$\Sigma = \int\limits_{h_1}^{h_2} \sigma \, dz = \begin{pmatrix} \Sigma_{xx} & \Sigma_{xy} \\ -\Sigma_{xy} & \Sigma_{yy} \end{pmatrix}. \tag{2.20}$$

Figure 2.6 shows typical values of Σ_{ij} versus latitude for daytime conditions. These conductivities also vary with time of day, season, and geomagnetic activity. The conductivities decrease by a factor of more than 10 during the night.

The total resistance of the dynamo region is of the order of

$$R_D \simeq \pi/\overline{\Sigma_{xx}} \simeq 30 \, \text{m}\Omega \, . \tag{2.21}$$

This value is four orders of magnitude smaller than the resistance R_A of the fair weather regions within the lower and middle atmosphere [Eq. (2.8)].

Since the average electric conductivity of the earth's surface is about $\sigma_E \simeq 10^{-3} \, \text{S/m}$, which is one order of magnitude larger than the maximum Pedersen conductivity in the ionosphere, the total resistance of the ground for quasi-static global electric currents can be estimated to be of the order of

$$R_E \simeq 1 \, \text{m}\Omega \, . \tag{2.22}$$

I Thunderstorms and Related Phenomena

3 Global Electric Circuit

In the following five chapters, we deal with electric fields and currents generated in thunderstorms. Thunderstorms behave like batteries which are connected with the highly conducting ionosphere and earth via the barely conducting lower and middle atmosphere (Fig. 3.1). The passive electric quasi-continuous current flowing outside the thunderstorm regions down to the earth is part of the global electric circuit. In this chapter, we outline some characteristics of the global electric circuit. In the following chapters, we discuss the thunderstorm as area of charge separation, lightning events as breakdown electric currents, and sferics as the electromagnetic radiation of the lightning currents.

3.1 Electric Fields and Currents

In fair weather regions far away from thunderstorm areas, one measures a downward-directed current density of the order of $j_z \simeq -2 \times 10^{-12}\,\text{A/m}^2$, which is believed to be driven by the global thunderstorm activity. This current density is remarkably constant with height (Fig. 3.2). Since the air is electrically conducting, the current is accompanied by a downward-directed electric field of the order of $E_z \simeq -100\,\text{V/m}$ on the ground. Ohm's law demands, then, that

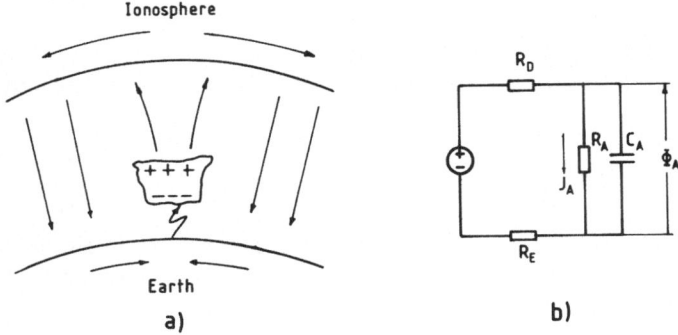

a)   b)

Fig. 3.1. a Schematic view of global electric circuit between earth and ionosphere with thunderstorms as generator. **b** Equivalent electric circuit. The thunderstorm generator is discussed in Chap. 4

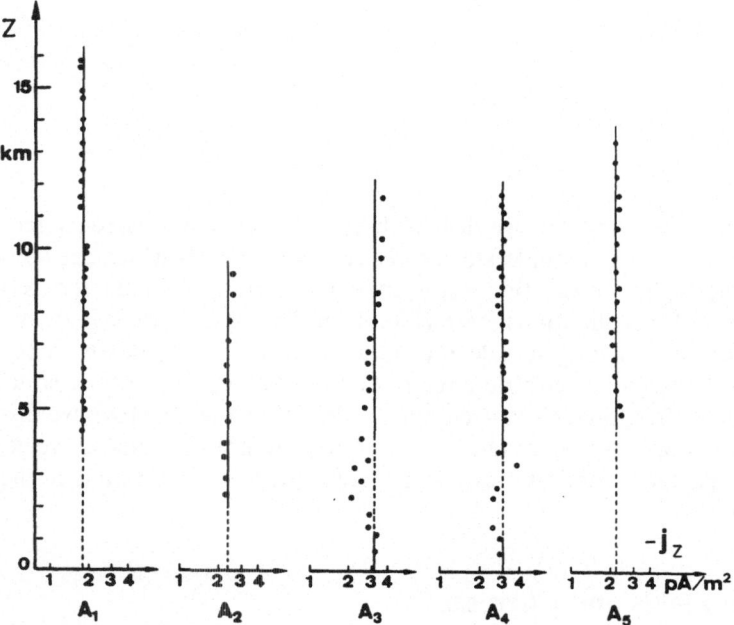

Fig. 3.2. Balloon measurements of vertical electric current density at fair weather during fall 1973 over the North Atlantic. (Gringel et al. 1978)

$$E_z = \varrho j_z \, , \tag{3.1}$$

and since $j_z \simeq$ const, $|E_z|$ must decrease with height like the specific resistance ϱ [Eq. (2.4); see also Fig. 2.2]. This yields

$$E_z = -\sum_{i=1}^{3} E_i \exp(-\alpha_i z) \, , \tag{3.2}$$

with $E_i = |j_z|\varrho_i$ (see Table 2.1).

Integration over the height gives the electric potential Φ between ground and a height z:

$$\Phi = -\int_0^z E_z \, dz = -\Gamma j_z = \sum_{i=1}^{3} \Phi_i [1 - \exp(-\alpha_i z)] \, , \tag{3.3}$$

with Γ from Eq. (2.6) and $\Phi_i = |j_z|\Gamma_i$ from Table 2.1. Here, zero potential is defined to be at the ground [$\Phi(0) = 0$]. The average potential between ground and ionosphere, called the atmospheric electric potential, is

$$\Phi_A = -\Gamma_A j_z \simeq 240 \, \text{kV} \, , \tag{3.4}$$

varying between about 180 and 400 kV [Mühleisen 1971; Markson 1981], and the total electric current in the fair weather regions is

$$J_A = -\Phi_A/R_A \simeq -1 \, \text{kA} \, . \tag{3.5}$$

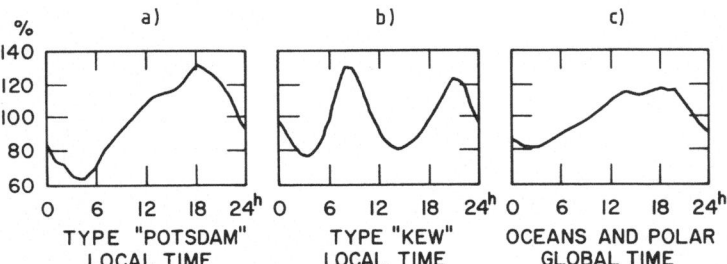

Fig. 3.3a – c. The three main types of diurnal variation of the vertical electric fair weather field on the ground. Types **a** and **b** are continental types depending on local time. Type **c** is an oceanic type depending on universal time. (Israel 1973)

The quasi-static electric field on the earth's surface is always orthogonal to the ground. This indicates that the earth's surface behaves like an electric equipotential layer for quasi-continuous electric fields.

At ocean and polar stations, one measures on the average a universal time (UT) dependence of E_z and j_z with maximum amplitudes near 1800 GMT [type (c) in Fig. 3.3]. The average columnar resistance Γ_A over these stations is nearly constant so that this UT dependence reflects the temporal variation of the source of Φ_A, which is believed to be thunderstorm activity (see Chap. 4).

The columnar resistance in mountain areas is smaller than at sea level. Therefore, a constant potential Φ_A at a given time will produce a greater electric current density in mountain areas than at sea level.

At continental stations, one observes on the average two types of local time dependence in E_z, a diurnal variation and a semidiurnal variation. The diurnal variation of type (a) in Fig. 3.3 occurs predominantly during winter at many stations, whereas the semidiurnal variation of type (b) generally dominates during summer. However, some stations, mainly in cities, measure type (b) variations throughout the year, and others, predominantly in rural areas, have only the diurnal variation of type (a). Type (b) is limited to plains and disappears above several tens of meters above ground where a clearly marked transition to type (a) variations is observed.

The local-time dependence of E_z at continental stations is remarkably similar to the variation of the water vapor pressure. The type (a) variation of water vapor pressure is called the maritime diurnal variation and occurs whenever the ground provides a continuous and adequate supply of water vapor by evaporation.

The water vapor pressure will then approximately follow the diurnal temperature curve, thus changing the ratio of small to large ions, and therefore the columnar resistance within the atmospheric boundary layer. The type (b) variation of the water vapor pressure stems from the diurnal component with the additional effect of convection. Convection with its maximum at noon can reduce the water vapor content near the ground during summer, thus overcompensating for the evaporation effect during daytime (Israel 1973). Days

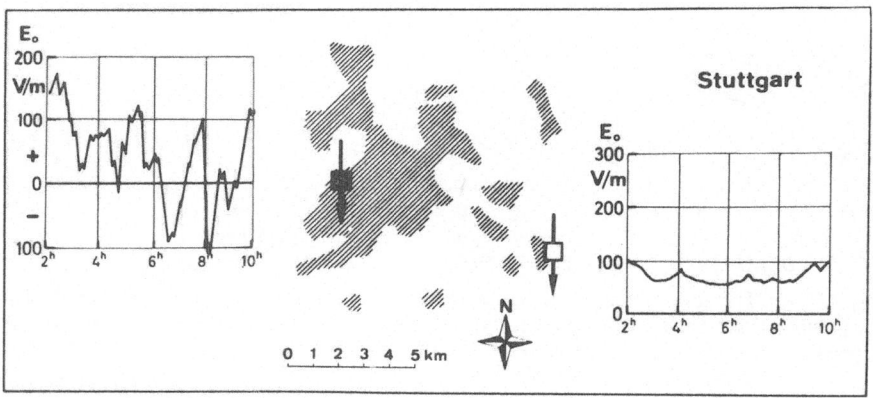

Fig. 3.4. Anthropogenic influence on the vertical electric field on the ground. Simultaneous measurements at two stations east and west of Stuttgart, Germany. The station west of the town is situated in an industrial area, while the station east of the town lies in a rural area and dispays a fair weather field (Fischer 1977). (Coordinate system of Fig. 14.1b is adopted)

with an electric field behavior as in Fig. 3.3 are called fair weather days in a geoelectric sense.

Fog, haze, or rain, but also man-made pollution, can substantially alter the electric conductivity near the ground. Simultaneous individual records of the field strength may, therefore, greatly differ even for nearby stations. Figure 3.4 shows anthropogenic influences on the ground-level electric field within an industrial area (left), whereas the station in a nearby rural area (right) exhibits a fair weather field.

Large volcano eruptions that expel ash and gas into the atmosphere can drastically increase the aerosol content of the middle atmosphere for years. The electric conductivity decreases, and the global atmospheric potential increases during these events (Hofmann and Rosen 1977; Meyerott et al. 1983).

Enhanced solar cosmic ray intensity during major solar flare events can increase the electric conductivity above about 15 km height for hours and can even produce a transient reversal of the vertical electric field within the middle atmosphere (Reagan et al. 1983). Sounding rockets have recently detected large vertical electric fields [of the order of a few V/m; many orders of magnitude larger than the normal field (see Fig. 2.2 and Eq. (3.2))] in the height range near 65 km (Maynard et al. 1981). The origin of these fields is not known, and the measurements are still controversial (Kelley et al. 1983).

3.2 Space Charges

Static electric fields are produced by electric charges. The observed decrease of the vertical electric field with height is necessarily accompanied by electric space charges. The distribution of these space charges follows from Coulomb's equation [Eq. (14.13)] as

$$q = \varepsilon \partial E_z / \partial z \qquad (3.6)$$

where $\varepsilon \simeq \varepsilon_0$ is the permittivity of the air. With the assumed electric field distribution in Eq. (3.2), this yields a positive space charge of

$$q = \sum_{i=1}^{3} q_i \exp(-\alpha_i z), \qquad (3.7)$$

with $q_i = \varepsilon \alpha_i E_i$ (Table 2.1 and Fig. 2.2).

The space charge at sea level becomes

$$q_A = 3.92 \times 10^{-12} \, \text{C/m}^3. \qquad (3.8)$$

This corresponds to an excess of singly charged positive small ions at sea level of

$$n_+ - n_- = q_A/e = 2.5 \times 10^7 \, \text{m}^{-3}, \qquad (3.9)$$

which is, in fact, small compared with the density of the fast ions on the ground of $n \simeq 10^9 \, \text{m}^{-3}$ (see Sect. 2.2).

The total columnar space charge within the atmosphere is

$$\hat{q}_A = \int_0^\infty q \, dz = \sum_{i=1}^{3} q_i/\alpha_i = 1.33 \times 10^{-9} \, \text{C/m}^2, \qquad (3.10)$$

which gives a total charge in the entire atmosphere of

$$Q_A = \hat{q}_A F_E = 677 \, \text{kC}. \qquad (3.11)$$

The total space charge induced on the earth's surface is then $-Q_A$, in order to compensate for this atmospheric space charge.

The electric conductivity is reduced in local dust and fog so that enhanced layers of space charge with horizontal gradients may form. One source of space charge near the ground is due to the electrode effect (Anderson 1977). The negative ions drift upward and the positive ions drift downward under the influence of the vertical electric field. Since the ions are formed only within the air, a depletion of the negative ions occurs near the ground, giving rise to an enhancement of positive space charge up to a factor of 100 within the first 10 m above ground. This effect is normally absent above land masses because of the strong vertical turbulent mass exchange. Negative ions emanating from a waterfall, or positive ions created by breaking surf in a salty ocean, fire, air pollution, or radioactive material in building construction are further sources of local space charge variations (Dolezalek 1972; Reiter 1980).

3.3 Equivalent Global Electric Circuit

We can simulate the passive global electric system between earth and iono-sphere by an equivalent RC circuit (Fig. 3.1), which has a resistance R_A [Eq. (2.8)], a capacitance of

$$C_A = Q_A / \Phi_A \simeq 2.9 \, \text{F} \qquad (3.12)$$

and a time constant of

$$\tau_A = R_A C_A \simeq 11 \, \text{min} . \qquad (3.13)$$

The system would discharge within a few tens of minutes if the charge generators (the thunderstorms) were to cease operating.

4 Thunderstorms

Thunderstorms are the main source of electromagnetic energy within the lower atmosphere and are believed to drive the global electric circuit. In this chapter, we summarize our basic knowledge about the phenomenon of thunderstorms.

4.1 Occurrence of Thunderclouds

Typical thunderclouds are convective cumulo-nimbus clouds with vigorous updrafts and downdrafts. Because heating from the sun is the primary source of the formation of temperature gradients in the atmosphere, strong thunderstorms are most common in tropical and temperate zones. Thunderstorm activity is most intense and frequent during the spring and summer, with maximum intensity over land at midafternoon (Vonnegut 1982).

Most, but not all convective clouds accumulate a net positive electric charge in their upper and a net negative charge in their lower regions. The electric fields from these charges become sufficiently intense to reverse the fair weather field over and beneath the clouds and to generate electric currents that can maintain the negative charge of the earth against the fair weather field (Fig. 3.1).

Thunderstorms tend to occur in crowds. Severe storms are often orientated in lines of 100 and more km in length (squall lines). Each thunderstorm contains several cells. The mean lifetime of one active thunderstorm cell is of the order of 30 min.

About 2000 thunderstorms are active around the earth at all times. They cover an area of about $5 \times 10^{11} \, \text{m}^2$, or 1/1000 of the earth's surface (Kasemir 1959). Figure 4.1 shows the annual mean of the area of thunderstorm activity versus universal time (UT) of the three main source regions in Africa/Europe, Asia, and America. Although the maximum intensity over the respective areas is at local midafternoon, the global activity displays a dependence on universal time with a maximum near 1800 h GMT. This result is a strong argument in favor of the Wilson theory, which attributes the global electric field to the thunderstorm activity (e.g., Israel 1973). Both curves peak at the same UT. However, there remain some discrepancies. The electric field in Fig. 3.3c, for example, varies only by 20%, while the thunderstorm area in Fig. 4.1

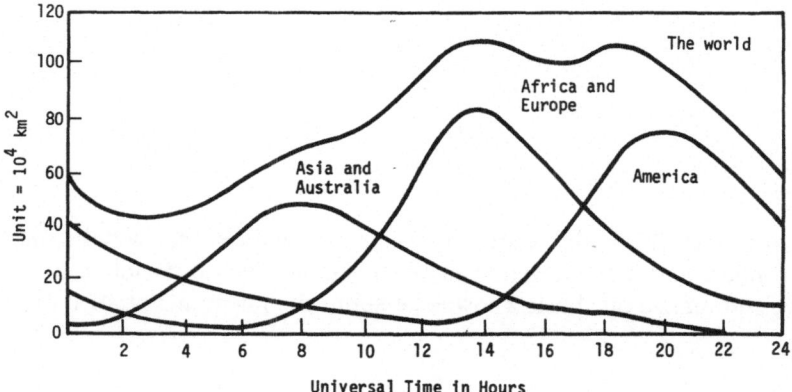

Fig. 4.1. Annual mean of diurnal variation of the area of glocal thunderstorm activity in the three main activity centers of the world. (Israel 1973)

varies by about 40%. Moreover, less convincing correlations are found if one splits the thunderstorm data into seasonal or monthly curves (Ogawa et al. 1969). One explanation for the discrepancy may be the still poor reliability of the thunderstorm data which are based even today on audible and visible observations ("a thunderstorm day", according to meteorological definition, "is a day where thunder has been heard").

4.2 Development of a Thunderstorm Cell

A parcel of air is convectively instable or buoyant and rises if the temperature–height profile decreases more strongly than adiabatically. The adiabatic temperature decrease with height (the dry adiabatic lapse rate) which a parcel of dry air follows is

$$dT/dz = -g/c_p = -\Gamma \simeq -0.01 \text{ K/m} , \tag{4.1}$$

with $g = 9.81 \text{ m/s}^2$ the gravitational acceleration, and $c_p \simeq 1000 \text{ J/(K kg)}$ the specific heat at constant pressure. Water vapor, which is always present, partly condenses to the liquid phase if the ascending and cooling moist air becomes saturated. The latent heat released during the phase transition from water vapor to liquid water acts to increase the temperature of the air parcel, so that the parcel now follows a moist adiabatic lapse rate of $\Gamma_m < \Gamma$ during its ascent. If the temperature is below freezing, transformation to ice occurs with the additional release of latent heat. Since freezing only occurs on appropriate nuclei, most clouds remain liquid until the temperature falls below $-10°$ to $-20°$C.

In a developing isolated thunderstorm cell the motion is primarily upward as moist air from the sub cloud boundary layer rises above the condensation

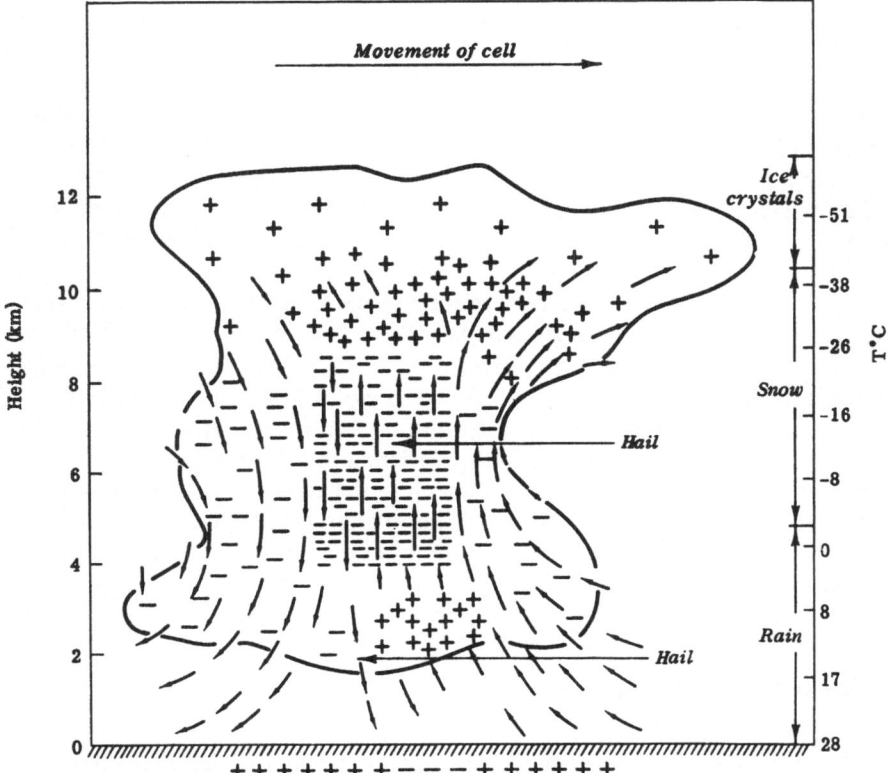

Fig. 4.2. Typical thundercloud cell illustrating its dimension, electric charge distribution, and convective winds

level where it becomes buoyant. This updraft causes a cloud of several km thickness (Fig. 4.2). The water drops formed during this updraft can be divided into two types: small drops with typical radii of 10 to 100 μm which remain suspended in the air and are carried upward by the updraft (cloud drops), and rain drops large enough to have fall velocities equal to or greater than that of the updraft (5 – 10 m/s). These drops (rain water) reach typical radii of 0.1 to 1 mm and have a water content of typically 0.1 to 1 g per kilogram of air.

Since the temperature in the top of the clouds may reach −50 °C, many precipitating particles are in form of snow crystals. The shapes of these crystals are modified as they fall into lower warmer regions of the cloud. They may aggregate to snowflakes as large as 3 cm in diameter. As they fall through relatively warm regions of the cloud, the snow crystals encounter supercooled cloud droplets which are captured and freeze on the surface of the crystals (riming). If the crystals become heavily rimed, the resultant particles are called graupel. Below the 0 °C level in the cloud, the frozen particles begin to melt and fall to the ground as rain. If the falling rain drops then encounter colder air (for example, in a temperature inversion), they may reach the ground as small pellets of ice.

When the rain water falls through the middle level of the cloud, the weight of this water increases the effective air density so much that it reduces and may even eliminate the air buoyancy in the lower half of the cloud. Meanwhile, the upper part of the updraft, losing its water content along the way, reaches a level at which it is no longer buoyant. This level is usually near the tropopause at about 10 to 15 km altitude. The cell has now reached its mature stage.

The accumulated rain water falls rapidly to the ground as a heavy shower accompanied by a cold evaporating downdraft. The air in this downdraft is much colder than that which originally rose in the updraft because the rain continues to cool it by evaporation below the cloud base. The thundercloud cell finally dissipates because these downdraft winds cut off the supply of warm moist air feeding the updraft in the original cell. At the same time, this surge of cold air is capable of lifting moist air in the adjacent environment and thereby triggering new cell formation. The process may repeat itself several times so that the total thunderstorm lifetime is typically about 1 to 2 h (Lilly 1979; Magono 1980).

4.3 Rain Drop Spectra

Rain drops have spherical shapes if their radii are smaller than about 150 μm. They are slightly deformed at radii between 150 μm and 500 μm, and they break up at radii larger than 5 mm. In the case of nonspherical rain drops, one can define an equivalent radius corresponding to a sphere of equal volume. Typical spectra of rain drops during the active phase of a storm can be described by the distribution (Marshall and Palmer 1948)

$$W(r) = (2\bar{n}/\bar{r}) \exp(1 - 2r/\bar{r}) \quad (\text{for } r \geqq \bar{r}/2) , \tag{4.2}$$

where W is the probability of the drops to have the equivalent radius r.

The total number of drops per volume is

$$\bar{n} = \int_{\bar{r}/2}^{\infty} W(r)\, dr . \tag{4.3}$$

The average equivalent radius is

$$\bar{r} = (1/\bar{n}) \int_{\bar{r}/2}^{\infty} r\, W(r)\, dr . \tag{4.4}$$

The water content per volume becomes

$$\varrho_c = (4/3)\, \pi \varrho_W \int_{\bar{r}/2}^{\infty} r^3\, W(r)\, dr = 8 \pi \varrho_W \bar{n} \bar{r}^3/3 \tag{4.5}$$

with $\varrho_W = 10^3 \, \text{kg/m}^3$ the density of water. The terminal fall velocity of the drops is

$$w_\infty \simeq -Gr \qquad (4.6)$$

with $G \simeq 1 \times 10^4 \, \text{s}^{-1}$ (Pruppacher and Klett 1980). The rainfall rate at terminal velocity is then

$$K = (4/3) \, \pi G \int\limits_{\bar{r}/2}^{\infty} r^4 W(r) \, dr = 65 \, \pi \bar{n} G \bar{r}^4/12 \qquad (4.7)$$

or $\Lambda = 0.65 \, (n_0/K)^{0.2}$ with $\Lambda = 1/\bar{r}$ (in mm^{-1}), $n_0 = 5.44 \, \bar{n} \Lambda$ (in m^{-3} mm^{-1}), and K (in mm/h). Λ and n_0 are the original symbols in the Marshall-Palmer formula.

As an example, if one assumes $\bar{r} = 0.5$ mm and $\bar{n} = 200 \, \text{m}^{-3}$, one arrives at $\varrho_c = 2 \times 10^{-4} \, \text{kg/m}^3$, and $K = 2.1 \times 10^{-6} \, \text{m/s}$, corresponding to a rainfall rate of 7.6 mm/h.

4.4 Electric Fields and Charges in Thunderclouds

Measurements of the electric field in clouds and their surroundings indicate that most thunderclouds behave to a first approximation like a huge dipole with positive electric charge of the order of 40 Coulomb concentrated in the top region above the $-20\,°C$ level, and negative charge of the same order accumulated in the center between the $-20\,°C$ and the $0\,°C$ level of the cloud. A smaller local positive charge region of the order of 10 Coulomb is sometimes observed below the $0\,°C$ level (Fig. 4.2).

Typical electric fields in and beneath a thundercloud are shown in Fig. 4.3. One usually observes precipitation beneath a developing cloud prior to significant electric perturbation. The initial electrification of a cloud as measured beneath the cloud is usually a reversal of the fair weather field polarity. When the field exceeds about 2 kV/m, it is enhanced locally over well-exposed grounded objects with small radius of curvature such as branches of trees, bushes, grassland, or man-made objects like rod antennas, to such an extent that point-discharge ions with positive charge are released into the air. Light rain arriving at the earth at this time frequently carries a positive charge which may have been captured from the positive point-discharge ions. After a period of about 10 min, the field strength in the cloud may have reached values up to $-400 \, \text{kV/m}$, while the field strength on the ground caused by the cloud overhead increases to values of the order of $+5 \, \text{kV/m}$. Electric breakdown and lightning flashes can now occur in the cloud. Many clouds, however, never reach the lightning stage even though they develop precipitation and strong electric fields. Coincident with the discharge, the surface electric field makes a large discontinuous excursion back to the fair weather field polarity because negative charge has been abruptly removed from overhead by the flash.

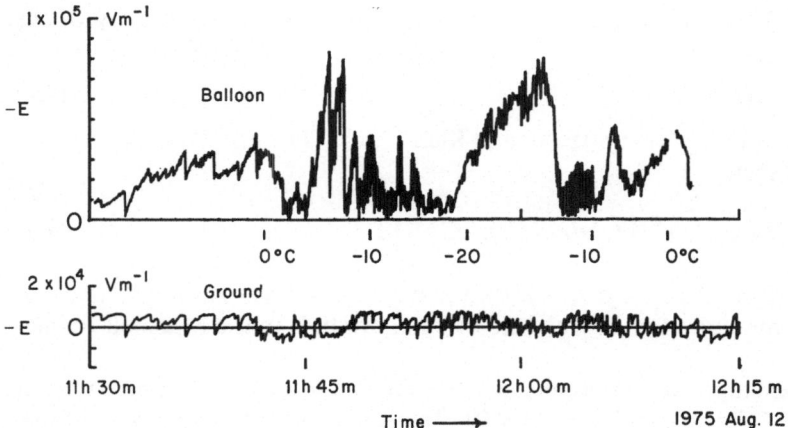

Fig. 4.3. Simultaneous records of vertical electric field from a mountain top (*lower panel*) and from a balloon rising into a thundercloud over the mountain top (*upper panel*). The temperature indicates the height of the balloon. Note that the direction of the electric field is defined as positive downward (Fig. 14.1b). (Winn and Byerley 1976)

The field strength in the cloud breaks down during a discharge. However, there is normally no change in polarity. The recovery after a discharge occurs within several tens of seconds, after which a new flash may start. Often, a gush of rain follows a lightning flash. The field above the cloud is generally directed upward and is accompanied by an upward directed electric current of the order of 1 ampere integrated over the total area of the thunderstorm (Vonnegut et al. 1966; Moore and Vonnegut 1977).

Observations show that the electric charge in the clouds is carried mainly by the droplets and the ice crystals, the small cloud drops carrying predominantly positive charge, the large rain drops carrying mainly negative charge. The amount of charge per drop Q depends on their size (Fig. 4.4). Drops with a diameter of 1 mm can carry as many as $Q/e \simeq 2 \times 10^7$ elementary charges.

Water drops are dielectric and thus become polarized in an external electric field such that the secondary electric field of the polarized particles weakens the primary field. The upper limit of the charge a drop can carry is reached when the secondary field equals the primary field E in magnitude. This can be estimated from Coulomb's law as

$$Q_{\max} = 4\pi E \varepsilon r^2 \lesssim A r^2, \qquad (4.8)$$

with $E \lesssim E_{\max} \simeq 4 \times 10^5$ V/m, $E_{\max}$ the maximum electric field observed, r the equivalent drop radius, and $A = 4.4 \times 10^{-5}$ C/m^2. The line labeled "2" in Fig. 4.4, defining the limit given by Eq. (4.8), lies well above most of the observed electric charges of the drops. According to Eqs. (4.2) and (4.8), the average charge per drop is

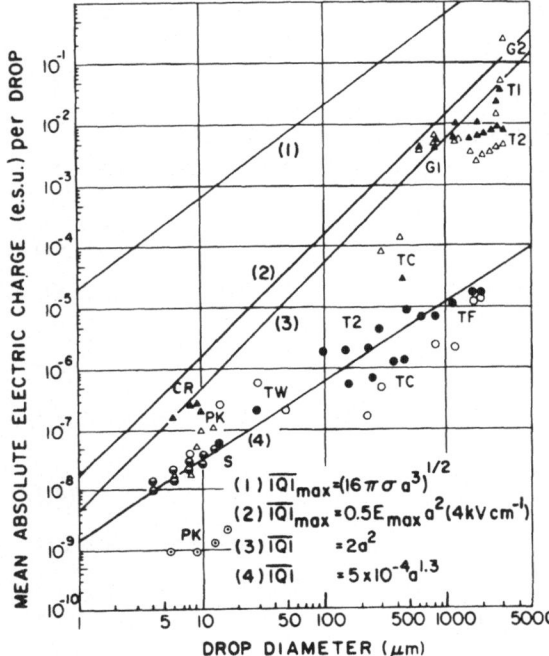

Fig. 4.4. Mean absolute electric charge on cloud drops and rain drops according to different authors. *Round symbols* indicate warm cloud cases, *triangular symbols* indicate thunderstorm cases, *solid symbols* indicate negative charge, *open symbols* indicate positive charge. (Pruppacher and Klett 1980). *Solid lines* are theoretical curves. Q_{max} maximum charge per drop; a equivalent drop radius; σ surface tension of water against air; E_{max} maximum electric field. Note that 1 esu corresponds to $(1/3) \times 10^{-9}$ C

$$Q \lesssim (A/\bar{n}) \int_{\bar{r}/2}^{\infty} r^2 W(r)\, dr = 1.25\, A\, \bar{r}^2. \tag{4.9}$$

The electric conductivity in the clouds is controled by the local balance of sources and sinks. The dominant sink is caused by cloud and aerosol particles, which reduce the electric conductivity within the clouds by a factor of 10 or more compared with the fair weather value because the small ions tend to become attached to the cloud particles. The conductivity decreases with increasing electric field. However, when the field strength in the cloud is very large ($|E| > 300$ kV/m), corona discharge from water droplets or ice pellets may greatly enhance the ion production rate and thus locally compensate for the decrease of the conductivity (Pruppacher and Klett 1980).

The difference in conductivity between clear air and a cloud causes a layer of space charge to form on the boundary between cloud and clear air if electric currents flow through that boundary. This surface charge produces a discontinuity in the electric field at the boundary. For an estimate, we simulate a cloud by a sphere of radius R, and with internal electric conductivity σ_i, embedded in an environment with conductivity σ_a. We assume that a radial electric current flows out of that sphere. This current must be continuous across the boundary at R:

$$j = \sigma_i E_i = \sigma_a E_a, \tag{4.10}$$

with the subscripts *"i"* and *"a"* referring to values inside and outside the boundary, respectively. The surface charge density on the sphere, according to Gauss' law and Eq. (4.10), is

$$\hat{q}_s = \varepsilon(E_a - E_i) = -\varepsilon E_i(1 - \sigma_i/\sigma_a) \ . \tag{4.11}$$

This surface charge density partly screens the internal charge Q_i.

The electric field on the sphere is related to the internal charge Q_i according to Coulomb's law as

$$E_i = Q_i/(4\pi\varepsilon R^2) \ . \tag{4.12}$$

Integration over the surface of the sphere relates the total surface charge Q_s to the charge inside the sphere:

$$Q_s = \int \hat{q}_s dF = -Q_i(1 - \sigma_i/\sigma_a) \ . \tag{4.13}$$

The effective charge of the sphere as measured by an observer outside the sphere is therefore only a fraction of Q_i:

$$Q_{\mathrm{eff}} = Q_s + Q_i = (\sigma_i/\sigma_a) Q_i \ . \tag{4.14}$$

Since an electric current flows out of the sphere, the sphere behaves like a current generator of strength

$$J_p = \int j \, dF = (\sigma_i/\varepsilon) Q_i = (\sigma_a/\varepsilon) Q_{\mathrm{eff}} \ . \tag{4.15}$$

The factor

$$\tau = \varepsilon/\sigma_a \tag{4.16}$$

is a time constant of the clear air at that height [see also Eq. (3.13)].

5 Thunderstorm Electrification

The typical characteristics of thunderclouds is separation of electric charge against the fair weather field to such an extent that breakdown electric currents (lightning) flow. In this chapter, basic concepts of the electrification mechanism in thunderclouds are outlined and simple models of the electric field and current configurations in the environment of the clouds are presented.

5.1 Charge Separation in Clouds

Positive and negative electric charges in a thundercloud are gravitationally separated against the fair weather field by updraft winds so that positive charge is deposited near the top and negative charge near the center of the clouds. Discharging outside the cloud occurs mainly via the electrically conducting lower atmosphere-ionosphere-earth system known as the global electric circuit (Fig. 3.1). Inside the cloud, lightning discharges act like short circuits.

Although several theories of charge separation exist, none of them is generally accepted (e.g. Mason 1971; Magono 1980). Most of these theories are based on one of two primary concepts: either inductive processes or noninductive processes, both of which preassume precipitation as the power mechanism.

In the inductive process, an external electric field induces electric polarization of the rain drops or ice pellets, the degree of polarization depending on the dielectric constant of the particles involved. Small ice crystals or water droplets colliding with these larger particles may acquire positive charge which they can carry upward within the updraft. The negatively charged rain drops and ice pellets descend due to their greater weight, thus enhancing the original electric field (Fig. 5.1). The amount of charge exchange between the large and small particles increases with increasing electric field. The effect is thus supported by a positive feedback that increases the original electric field until either the limiting charge value of Q_{max} [Eq. (4.8)] is reached, accompanied by lightning, or the gravitational force is offset by the electric force, thereby stopping the descent of the larger particles.

Laboratory experiments have shown that the inductive process becomes significant only if relatively strong fields of the order of 10 kV/m or more are

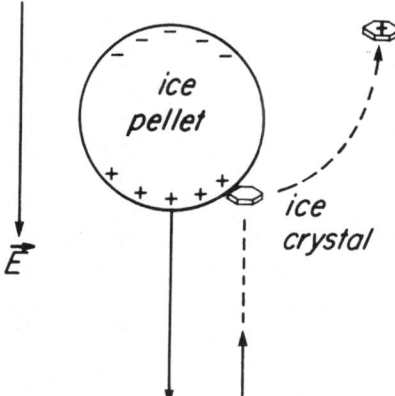

Fig. 5.1. Schematic view of inductive charge separation. A falling ice pellet is polarized by the external electric field E. An ice crystal colliding with the pellet acquires positive charge and is carried upward by the updraft wind. The heavier negatively charged pellet falls down. (Iribarne and Cho 1980)

already present (Aufdermauer and Johnson 1972). Noninductive processes are, therefore, also expected to be important. Among others, these processes may be of thermoelectric nature. If two pieces of ice initially at different temperatures are brought together and then separated, the warmer piece acquires negative charge while the colder one acquires an equal amount of positive charge. This is due to the diffusion of the more mobile positively charged hydrogen ions down the temperature gradient, while the more immobile negatively charged OH^- ions remain in excess in the warmer portion. Since ice crystals and graupel are often present under conditions of strong cloud electrification, and since the graupel particles are generally slightly warmer than the environment due to the latent heat release from accreted supercooled drops, collisions between small ice crystals and the graupel particles may favor thermoelectric charging.

Grenet (1947) and Vonnegut (1953) proposed a convection process which does not depend on precipitation. The idea is that the positive space charge which is usually present in the fair weather regions is transported by updraft into the cloud. Here the ions become attached to the cloud droplets, leaving a net positive charge there and decreasing the electric conductivity. The electric field of this charge can cause negative ions in the clear air around and above the cloud to migrate to the cloud surface where they become trapped on cloud particles, causing the external portions of the cloud to have negative charge. Vigorous updraft inside the cloud and corresponding downdraft motions at the outer portions of the cloud transport the positive charges to the top and the negative charges to the lower levels of the cloud.

Another charging mechanism not invoking precipitation is based on the experimental observation that the free electrons present in the air are mainly attached to the NO_2 and the O_2 molecules (Wahlin 1977). When the neutral cloud droplets are ventilated by the air, these negatively charged ions react with the surface of the cloud particles, transferring negative charge to them.

The positive ions are left in the surrounding air, building up electric double layers. A quantitative model based on this electrochemical mechanism is due to Pathak et al. (1980). Since occasional lightning events can be observed without significant precipitation, it is apparent that these charge separation mechanisms must be at work in some cases.

5.2 Theory of Precipitation-Powered Electrification

In the following, we consider in more detail the two precipitation-generated processes which are presently favored in the literature (Mathal et al. 1980; Kuettner et al. 1981). We assume a simplified one-dimensional model and describe the growth of the (negative) charge Q_l of the large particles (graupel and/or rain drops) by the balance equation

$$\partial Q_l/\partial t = v(\hat{Q} + \beta E - \gamma Q_l), \qquad (5.1)$$

where $\hat{Q}$ is the negative charge transferred per collision to the large particles in the noninductive process, v is the frequency of collisions between one large particle and all small particles which may be ice crystals and/or cloud droplets, β is the efficiency of the inductive process, E is the vertical electric field, and γ describes loss processes of the charge of the large particles such as recombination, disruption, etc.

Kuettner et al. (1981) use the expressions

$$v = \pi r_l^2 w n_s \alpha; \quad \beta = (8/3)\pi \varepsilon \psi r_s^2,$$

where the following definitions are introduced:

r_l, r_s: equivalent average radii of the large and small particles, respectively
n_l, n_s: their number densities
w_l, w_s: their vertical velocities
Q_l, Q_s: their electric charges
$w = w_s - w_l > 0$: their relative velocity
ε: permittivity of the large particles
$\alpha \simeq 1$: efficiency factor of the collision process
$\psi \simeq \pi^2/2$: efficiency factor of the induction process

The downward transport of the negatively charged large particles and the upward transport of the positively charged small particles due to the combined action of updraft and gravitation corresponds to a charging electric current given by

$$j_c = n_s w_s Q_s + n_l w_l Q_l. \qquad (5.2)$$

This current is partly compensated by a dissipative conduction current

$$j_d = \sigma E, \qquad (5.3)$$

where σ is the electric conductivity within the cloud. A relationship between the electric field and current then follows from Ampere's law:

$$\varepsilon \partial E/\partial t + j_d + j_c = 0. \qquad (5.4)$$

We assume homogeneous conditions and quasi-neutrality:

$$q = n_s Q_s + n_l Q_l \simeq 0 . \tag{5.5}$$

Upon elimination of Q_l, Eqs. (5.1) to (5.4) yield a differential equation for the vertical component of the electric field:

$$\varepsilon \partial^2 E/\partial t^2 + (\sigma + v \gamma \varepsilon)\,\partial E/\partial t + v(\sigma \gamma - \beta n_l w) E = v n_l w \hat{Q} , \tag{5.6}$$

which has the solution

$$E = A_1 \exp(\delta_1 t) + A_2 \exp(-\delta_2 t) + \hat{E} \tag{5.7}$$

with

$$\hat{E} = n_l w \hat{Q}/(\sigma \gamma - \beta n_l w)$$

$$\left.\begin{matrix} \delta_1 \\ \delta_2 \end{matrix}\right\} = (1/2\varepsilon)\{[(\sigma - v \gamma \varepsilon)^2 + 4 \beta v \varepsilon n_l w]^{1/2} \mp (\sigma + v \gamma \varepsilon)\}$$

$$A_1 = [E_0(\delta_2 - \sigma/\varepsilon) - \delta_2 \hat{E}]/(\delta_1 + \delta_2)$$

$$A_2 = [E_0(\delta_1 + \sigma/\varepsilon) - \delta_1 \hat{E}]/(\delta_1 + \delta_2) .$$

The boundary conditions $E(t=0) = E_0$; $Q_l(t=0) = 0$ have been adopted here, and all coefficients in Eq. (5.6) have been taken as constant.

Figure 5.2 (curve c) shows the electric field calculated from Eq. (5.7) based on the following numerical values:

$$\hat{Q} = -1 \times 10^{-15}\,\mathrm{C}; \quad r_l = 0.5\,\mathrm{mm}; \quad r_s = 20\,\mu\mathrm{m}; \quad n_l = 200\,\mathrm{m}^{-3};$$

$$n_s = 5 \times 10^5\,\mathrm{m}^{-3}; \quad w = 10\,\mathrm{m/s}; \quad \varepsilon = \varepsilon_0; \quad \gamma = 5 \times 10^{-3};$$

$$\sigma = 2 \times 10^{-14}\,\mathrm{S/m}; \quad E_0 = -100\,\mathrm{V/m} .$$

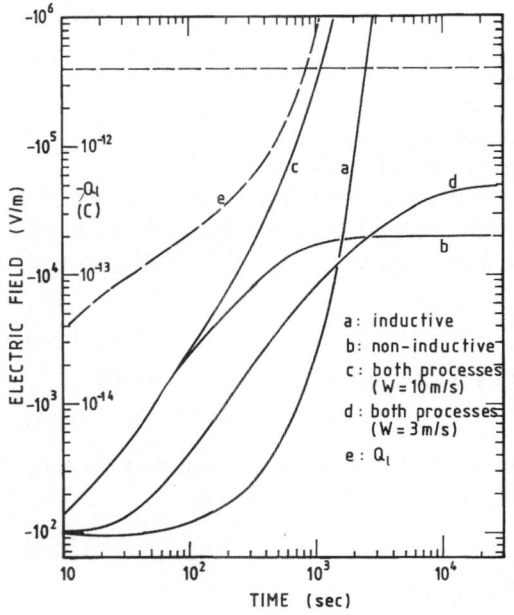

a: inductive
b: non-inductive
c: both processes (W = 10 m/s)
d: both processes (W = 3 m/s)
e: Q_l

Fig. 5.2. Calculated vertical electric field in a thundercloud versus time if the inductive mechanism (*curve a*), the non-inductive mechanism (*curve b*), or both processes (*curves c* and *d*) are effective. The negative electric charge versus time is plotted as *dashed line*

The electric field builds up from the fair weather field E_0 to the breakdown field of $-400\,\mathrm{kV/m}$ within about 20 min. The e-folding time of the exponential increase of E is

$$\tau = 1/\delta_1 \simeq (\sigma + v\gamma\varepsilon)/[v(\beta n_l w - \sigma\gamma)] \simeq 300\,\mathrm{s}\,. \qquad (5.8)$$

The e-folding time decreases with increasing relative velocity w.

If the charging current is too small to overcome the conduction current, the exponential growth stops. This happens if $\delta_1 < 0$, or if $\sigma\gamma > \beta n_l w$. As an example, curve d in Fig. 5.2 has been calculated with the same data as in curve c except that $w = 3\,\mathrm{m/s}$. The maximum field strength in this case is only $-50\,\mathrm{kV/m}$, too small to cause breakdown.

The noninductive process alone behaves similarly. In this case, one has $\beta = 0$, $\delta_1 = -\gamma v$, and $\hat{E} = n_l w \hat{Q}/(\sigma\gamma) \simeq -20\,\mathrm{kV/m}$ is the maximum field strength reached asymptotically (curve b in Fig. 5.2). On the other hand, the inductive process alone (curve a, calculated with $\hat{Q} = 0$) reaches breakdown fields after about 40 min, a build-up time twice as long as for the combined inductive and noninductive processes.

The growth of the electric charge Q_l can be determined from Eqs. (5.2). (5.4), and (5.5) as

$$Q_l = (\varepsilon \partial E/\partial t + \sigma E)/(n_l w)\,. \qquad (5.9)$$

Curve e (dashed line in Fig. 5.2) shows Q_l versus time with the numerical values used in curve c. Q_l reaches a value of $Q_{max} \simeq -10^{-11}\,\mathrm{C}$ at breakdown, a value within the limit given by Eq. (4.8).

One can estimate the upper limit of the total negative charge stored in the lower level of a thunderstorm as

$$Q_{total} \simeq n_l Q_{max} F \Delta h \simeq -500\,\mathrm{C}\,, \qquad (5.10)$$

where $F \simeq 250\,\mathrm{km}^2$ is the average area of one thunderstorm and $\Delta h \simeq 1\mathrm{km}$ is the typical thickness of the negative charge region. Considering the reduction of the internal conductivity compared with the fair weather conductivity [Eq. (4.14)] as well as the fact that a thunderstorm is composed of several cells, this number is not inconsistent with typical observed charges of $-40\,\mathrm{C}$ per thunderstorm cell.

The maximum charging current is

$$J_c = -n_l Q_{max} w F \simeq 5\,\mathrm{A}\,, \qquad (5.11)$$

while the maximum conduction current becomes

$$J_d = \sigma E_{max} F \simeq -2\,\mathrm{A}\,. \qquad (5.12)$$

The result of this simple estimate, in fact, shows that both processes must act together in order to generate the observed large electric fields and charges in a sufficiently short time and that breakdown fields are reached only if sufficiently large updrafts exist. More detailed calculations taking into account realistic winds including wind shears have been undertaken by Kuettner et al. (1981).

5.3 Thunderstorm as Local Generator

The electric charges in thunderclouds gain potential energy as they are mechanically separated against the existing electric field (electromotive force). Electric energy may succeed in recombining the charges of opposite sign. If this happens, we have an electric breakdown within the cloud, or an intracloud discharge. Such an event can be conceived as a short circuit of the electromotive generator.

The electrically conducting air allows a discharging current to flow inside and outside the cloud from the region of positive charge in the top to the region of negative charge in the lower level of the cloud. If the electric conductivity in the fair weather region were constant with altitude, that discharging current would flow mainly in the immediate environment of the cloud. However, since the conductivity increases nearly exponentially with height, a significant part of the discharging current flows from the top of the cloud into the ionosphere, thus contributing a current to the global electric circuit of about 0.5 ampere per thunderstorm.

We want to study in this section the streamlines of the discharging current outside a thunderstorm. The simplest model is a current point source simulating the upper positive charge center. Two point sources of opposite polarity are then used to approximate the observed quasi-dipole configuration of the cloud.

The electric potential of a current point source of strength J_p in steady state imbedded in a medium of constant conductivity σ is

$$\Phi_p = J_p/(4\pi\sigma r) \,, \tag{5.13}$$

where r is the distance between point source and observer, and the effective charge is related to the current strength J_p according to Eq. (4.15).

If the electric conductivity of the environment increases exponentially with height:

$$\sigma = \sigma_0 \exp(\alpha z) \tag{5.14}$$

the potential of the point source located at $z = 0$ changes to (Kasemir 1959)

$$\Phi_p = J_p \exp[-\alpha(r+z)/2]/(4\pi\sigma_0 r). \tag{5.15}$$

The electric current density which is determined by

$$j = -\sigma \nabla \Phi_p = \nabla \times \Psi \tag{5.16}$$

can be derived from a stream function Ψ, the azimuthal component of which (in a cylindrical coordinate system) is given by (Kasemir 1959)

$$\Psi^* = 2\pi\varrho\, \Psi_\lambda = \int_0^\varrho j_z \varrho'\, d\varrho' = J_p\{-(1/2)(1+z/r)\exp[-\alpha(r-z)/2]\}. \tag{5.17}$$

$\Psi^* = $ const. Determines a current streamline, and the numerical value of Ψ^* is the total current flowing within a circular area of horizontal radius ϱ at height z. The space charge can be determined from Eqs. (14.13) and (14.18) as (assuming steady state conditions)

$$q = \alpha \varepsilon \partial \Phi_p / \partial z = -J_p \varepsilon \alpha^2 / (8\pi\sigma_0 r)[1 + z/r + 2z/(\alpha r^2)] \exp[-\alpha(r+z)/2] \ . \tag{5.18}$$

If we place such a source at a height h above the earth's surface, simulated by a perfectly conducting plane at height $z = 0$, we must introduce a virtual current point sink of strength J_s located at height $-h$. For the case of constant conductivity of the air, the total potential of the point source with its image becomes simply

$$\Phi = \Phi_p + \Phi_s = (J_p/r_p + J_s/r_s)/(4\pi\sigma) \tag{5.19}$$

with

$$\left.\begin{array}{r} r_p \\ r_s \end{array}\right\} = [\varrho^2 + (z \mp h)^2]^{1/2} \quad \text{and} \quad J_s = -J_p \ .$$

Here, we have chosen the equipotential layer at $z = 0$ to be of zero potential. The strengths of the primary source and the secondary sink are equal.

For the more complicated case of an exponential increase of σ, the potential Eq. (5.19) changes to

$$\Phi = \{J_p \exp[-\alpha(r_p + z + h)/2]/r_p + J_s \exp[-\alpha(r_s + z - h)/2]/r_s\}/(4\pi\sigma_0) \ , \tag{5.20}$$

with

$$J_s = -J_p \exp(-\alpha h) = -J_p \sigma_0 / \sigma_h$$

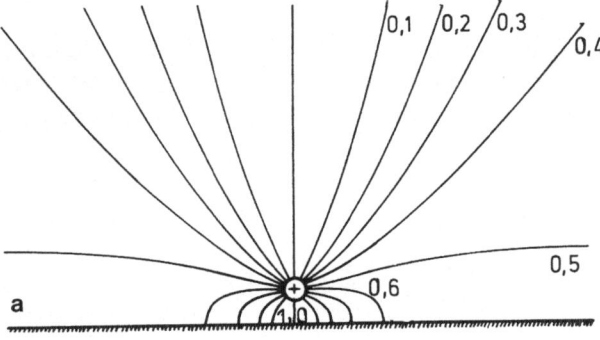

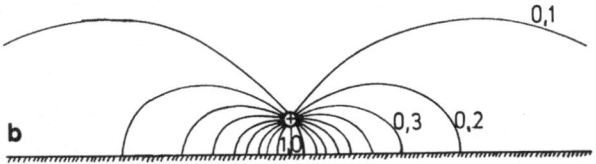

Fig. 5.3a, b. Current streamlines of a current point source embedded within an atmosphere with exponentially increasing conductivity over a perfectly conducting earth (*upper panel*), and the same point source embedded within an atmosphere of constant conductivity (*lower panel*). (Kasemir 1959)

and σ_0 and σ_h the conductivities at heights 0 and h, respectively. Corresponding formulae may be derived for the stream function and the space charge.

Figure 5.3b shows the current streamlines for the case of constant σ. Figure 5.3a is determined for a height-dependent σ according to Eq. (5.14) with $\sigma_h/\sigma_0 = 2$. Evidently, the virtual sink in Fig. 5.3a compensates only the fraction σ_0/σ_h of the primary source, while the remaining current J_c given by

$$J_c = (1 - \sigma_0/\sigma_h) J_p \tag{5.21}$$

flows into the upper atmosphere and contributes to the global electric circuit in the fair weather regions.

The efficiency of the current generator to produce the global electric circuit depends on the ratio between the electric conductivities at the height h and at the ground. The nearly exponential increase of σ with height is, therefore, essential for the existence of the global electric circuit. Moreover, the greater height of the positive charge in thunderclouds at low latitudes makes these thunderstorms more effective sources for the global electric circuit than the midlatitude thunderstorms.

In the more realistic case of a dipole generator, one must add the potentials of two point sources of equal strength but opposite polarity located at the heights h_1 and h_2. Figure 5.4 shows the current streamlines of such a dipole generator. The current between the two poles is a loss current within the cloud. It has the strength

$$J_d = -(J_p - J_c), \tag{5.22}$$

where $J_c = (\sigma_0/\sigma_1 - \sigma_0/\sigma_2) J_p$ is the fraction of the current flowing from the top of the cloud into the ionosphere and from the earth to the region of negative charge. σ_1 and σ_2 are the conductivities at the heights h_1 and h_2, respectively.

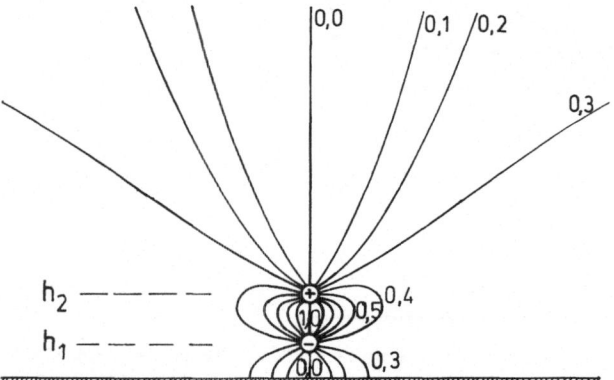

Fig. 5.4. Current streamlines from two point sources embedded within an atmosphere with exponentially increasing conductivity over a perfectly conduction earth. (Kasemir 1959)

The thunderstorm field, generating horizontal as well as vertical electric fields, modifies the strictly vertical global electric field in its environment. We can estimate the range where the thunderstorm field is still significant by comparing the horizontal component of the point source Eq. (5.20) with the vertical component of the fair weather field from Eq. (3.2):

$$|E_\varrho/E_{z0}| = J_p\varrho/(8\pi\sigma_0\,\Phi_A)\{[1+2/(\alpha r_p^2)]/r_p\exp[-\alpha(r_p+z-h)/2]$$
$$-[1+2/(\alpha r_s^2)]/r_s\exp[-\alpha(r_s+z-h)/2)]\}. \tag{5.23}$$

At the height of the source ($z = h \simeq 7.5$ km), this ratio decreases to unity at a distance of $\varrho = 40$ km, and to 0.01 at 75 km. The thunderstorm field at cloud heights is thus significant only within about 100 km distance from the source (Park and Dejnakarintra 1973).

5.4 Equivalent Current of Local Generator

The result of the last section suggests an equivalent current system of the form shown in Fig. 5.5, where J_c is the charging current flowing into the ionosphere and J_d is the lossy conduction current within the cloud. The circuit is closed via the fair weather areas simulated by the resistance $R_A \simeq 230\ \Omega$ [Eq. (2.8)] and the capacitance $C_a = C_A/N \simeq 1.5$ mF [Eq. (3.12)], with $N \simeq 2000$ the total number of active thunderstorms.

The resistance R_3 in region 3 (h_2, $h_3 \to \infty$) between cloud top and ionosphere is

$$R_3 \simeq 1/(\sigma_2 F) \int_{h_2}^{\infty} \exp(-\alpha z)\,dz = 1/(\sigma_0 F\alpha)\exp(-\alpha h_2). \tag{5.24}$$

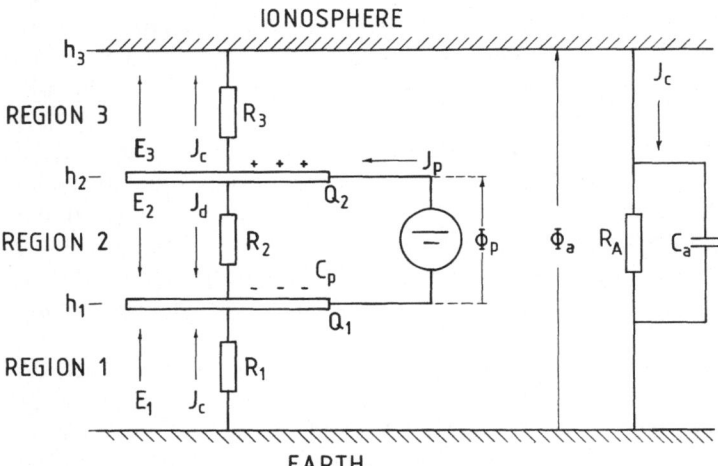

Fig. 5.5. Equivalent electric circuit of a thunderstorm generator

The internal resistance within the cloud [region 2; (h_1, h_2)] is

$$R_2 = (h_2 - h_1)/(\sigma_2 F) , \tag{5.25}$$

and the resistance between ground and cloud (region 1) becomes

$$R_1 = h_1/(\sigma_1 F) , \tag{5.26}$$

with σ_1 and σ_2 assumed to be constant. The regions of negative and positive charge are simulated by thin circular plates of area F located at heights h_1 and h_2, respectively. The average electric fields in regions 1 and 2 and the field at height h_2 above the upper plate are

$$E_1 = R_1 J_c/h_1; \quad E_2 = -R_2 J_d/(h_2 - h_1); \quad E_3 = R_3 J_c \alpha . \tag{5.27}$$

The electric charges on the plates are

$$Q_1 = \varepsilon F(E_2 - E_1); \quad Q_2 = \varepsilon F(E_3 - E_2) . \tag{5.28}$$

The currents are related by

$$\Phi_p = R_2 J_d = (R_1 + R_3 + R_A) J_c , \tag{5.29}$$

where Φ_p is the source potential and $J_p = J_c + J_d$ is the source current of the thunderstorm generator.

If we assume an average field strength in the cloud of $E_2 \simeq -100 \text{ kV/m}$, a charging current of $J_c \simeq 0.5 \text{ A}$; a discharging current of $J_d \simeq 0.2 \text{ A}$; furthermore, $\sigma_2 \simeq 1 \times 10^{-14} \text{ S/m}$; $\sigma_0 \simeq 5 \times 10^{-14} \text{ S/m}$; $1/\alpha \simeq 6 \text{ km}$ [σ_0, α approximate the conductivity profile at cloud heights and above; see Eq. (5.14)], $h_1 \simeq 5.5 \text{ km}$; $h_2 \simeq 7.5 \text{ km}$; $F \simeq 250 \text{ km}^2$, we arrive at $\Phi_p \simeq 200 \text{ MV}$; $J_p \simeq 0.7 \text{ A}$; $R_1 = 0.23 \text{ G}\Omega$; $R_2 = 1 \text{ G}\Omega$; $R_3 = 0.17 \text{ G}\Omega$; $Q_1 = -270 \text{ C}$; $Q_2 = 250 \text{ C}$; $E_1 = 21 \text{ kV/m}$; $E_3 = 14.3 \text{ kV/m}$; and $\sigma_1 = 1 \times 10^{-13} \text{ S/m}$. The capacitance of the thundercloud is $C_p \simeq Q_2/\Phi_p \simeq 1 \text{ μF}$. The current density below the cloud is $j_1 = J_c/F \simeq 2 \text{ nA/m}^2$. The electric energy of the generator is $W_p = \Phi_p J_p \tau = 1.4 \times 10^5 \text{ kWh}$ with $\tau \simeq 1 \text{ h}$, the average life time of a thunderstorm. This electric energy should be compared with the thermodynamic energy in the thunderstorm which is $W_t \simeq \varrho_c F \Delta h \lambda \simeq 5 \times 10^7 \text{ kWh}$, with $\varrho_c \simeq 2 \times 10^{-4} \text{ kg/m}^3$ the water content per volume of the cloud [Eq. (4.5)], $\lambda = 2.5 \times 10^6 \text{ J/kg}$ the latent heat of evaporation, and $\Delta h \simeq 2 \text{ km}$ a typical vertical extent of the cloud. Thus, the electric energy is only a small fraction ($<1\%$) of the thermodynamic energy. The relatively large numerical value of the conductivity σ_1 between cloud and ground is indicative of the increase in effective conductivity in that region due to coronal discharges, precipitation, cloud to ground lightning, and, possibly, displacement currents (Krider and Musser 1982).

In this example, the efficiency of the thunderstorm generator to generate a charging current is $|J_c/J_p| \simeq 0.71$. Each thunderstorm contributes to the fair weather current with about $J_c \simeq 0.5 \text{ A}$ and with a potential of $\Phi_a = J_c R_A \simeq 115 \text{ V}$. A total of N thunderstorms ($N = 2000$) is activated in parallel connection, thus increasing the total charging (or fair weather) current to $J_A = NJ_c \simeq 1$ kA and the total atmospheric potential to $\Phi_A = J_A R_A \simeq$ 230 kV. The respective resistances of the sum of all generators become $R_3/N \simeq 86$ kΩ; $R_2/N \simeq 0.5$ MΩ; $R_1/N \simeq 125$ kΩ. The total electric power is $P_A = NP_p \simeq 0.28$ TW. This power is only a minute fraction of the average solar heat input reaching the earth given by $(1 - A) SF_E/4 = 1.1 \times 10^5$ TW with $S = 1.37$ kW/m^2 the solar constant and $A \simeq 0.3$ the albedo.

5.5 Thunderstorm as Global Generator

In Sect. 5.3, we considered an individual thunderstrom in its immediate environment so that it was possible to apply a plane geometry without taking into account the finite size of the earth. However, we could not properly determine the closing circuit in the fair weather regions. That part of the circuit was introduced ad hoc in Fig. 5.5, based on plausible arguments about the closing current. Since the fair weather resistance R_A is small compared with the cloud resistances R_2 and R_3 in Fig. 5.5, no feedback is expected. The fair weather current in this plane model approaches zero because the plane is infinitely extended.

In this section, we derive solutions for the fields and currents of a thunderstorm in a spherical atmosphere. Since the resistance R_1 in Fig. 5.5 is small compared with R_2, we simplify the thunderstorm generator by setting $R_1 = 0$. This corresponds to a current point source of Eq. (5.15). If that point source is located at height $h = r - a$ above the earth's surface on the axis of a spherical coordinate system at $\theta = 0°$ (θ is the colatitude), we can represent the current source by

$$J = J_p \delta(z-h)\,\delta(x-1)/(\pi a^2) \tag{5.30}$$

with δ the delta function and $x = \cos\theta$. Furthermore,

$$\delta(x-1) = \sum_n (2n+1)P_n(x)/4 , \tag{5.31}$$

where $P_n(x)$ are the Legendre polynominals (e.g., Menzel 1960).

Again, we assume an electric conductivity which grows exponentially with height [Eq. (5.14)]. The electric potential can be determined from

$$\nabla \cdot j = - \nabla \cdot (\sigma \nabla \Phi) = J . \tag{5.32}$$

We develop Φ into a series of Legendre polynominals:

$$\Phi = \sum_n \Phi_n(z)P_n(x) , \tag{5.33}$$

and obtain a differential equation for $\Phi_n(z)$

$$(1/r^2)\partial/\partial r[r^2\partial\Phi_n/\partial r] + \alpha\partial\Phi_n/\partial r - n(n+1)\Phi_n/r^2 = 0 , \tag{5.34}$$

which has the general solution

$$\Phi_n(z) = A_n \exp(\beta_1 z) + B_n \exp(-\beta_2 z), \tag{5.35}$$

with

$$\left.\begin{array}{c} \beta_1 \\ \beta_2 \end{array}\right\} = \mp(1/a + \alpha/2) + [(1/a + \alpha/2)^2 + n(n+1)/a^2]^{1/2}.$$

$z = r - a \ll a$, and a the earth's radius.

If $n < 100$, it is $\beta_1 \simeq \gamma_n \alpha$ and $\beta_2 \simeq \alpha$ with $\gamma_n = n(n+1)/(\alpha a)^2 \ll 1$.

Taking into account the boundary conditions of continuous potential and currents, one arrives at

$$\Phi_n(z) = \begin{cases} A_n[\exp(\beta_1 z) - \exp(-\beta_2 z)] & \text{for} \quad \begin{array}{c} z < h \\ B_n \exp(\beta_1 z) + C_n \exp(-\beta_2 z) \end{array} \quad z > h \end{cases} \tag{5.36}$$

with

$$A_0 = R_A J_p; \quad B_0 = R_A J_c; \quad C_0 = 0 \quad \text{for} \quad n = 0$$

$$\left.\begin{array}{c} A_n = \\ C_n = \end{array}\right\} (2n+1) R_A J_p \alpha/(\beta_1 + \beta_2) \begin{cases} \exp[-(\alpha + \beta_1)h] \\ \exp[-(\alpha - \beta_2)h] - \exp[-(\alpha + \beta_1)h] \end{cases}$$

$$B_n = 0 \qquad\qquad\qquad\qquad\qquad\qquad\qquad\qquad \text{for} \quad n > 0$$

$$J_c = [1 - \exp(-\alpha h)] J_p = (1 - \sigma_0/\sigma_h) J_p$$

$$R_A = 1/(\sigma_0 \alpha F_E)$$

R_A is the total resistance in the fair weather regions [Eq. (2.8)]. The globally averaged atmospheric potential $(z \to \infty)$ is $\Phi_0 = B_0 = R_A J_c$.

With the values $\sigma_0 = 5 \times 10^{-14}$ S/m, $\alpha = 1/6$ km^{-1}, $J_p = 0.7$ A, $h = 7.5$ km, one obtains $J_c = 0.5$ A, $\Phi_0 = 118$ V, $R_A = 235$ Ω in fair agreement with the values in the preceding section.

If we select an arbitrary boundary at some height h_i (e.g., an electric wall with $\sigma \to \infty$ at 100 km), the globally averaged component ($n = 0$) would remain the same at altitudes $z < h_i$, while the higher order components ($0 < n < 100$) would become modified only within a few scale heights below the upper boundary. In other words, the upper region of the atmosphere has only a minor influence on the electric configuration within the lower and middle atmosphere.

More detailed calculations of the global and mesoscale structure of the thunderstorm field, taking into account orographic features of the earth as well as a realistic distribution of thunderclouds (simulated by dipole sources), have been made by Hays and Roble (1979). These authors assume an electric conductivity of the atmosphere which depends on height and latitude. Figure 5.6, taken from Hays and Roble, shows contours of the calculated potential at various constant surfaces of conductivity (which are nearly constant height levels). The thunderstorm regions are located between 8 and 15 km altitude. The potential difference $\Phi - \Phi_A$ in Fig. 5.6 (panels a to c) is positive above the main thunderstorm areas in Africa and America, indicating electric currents

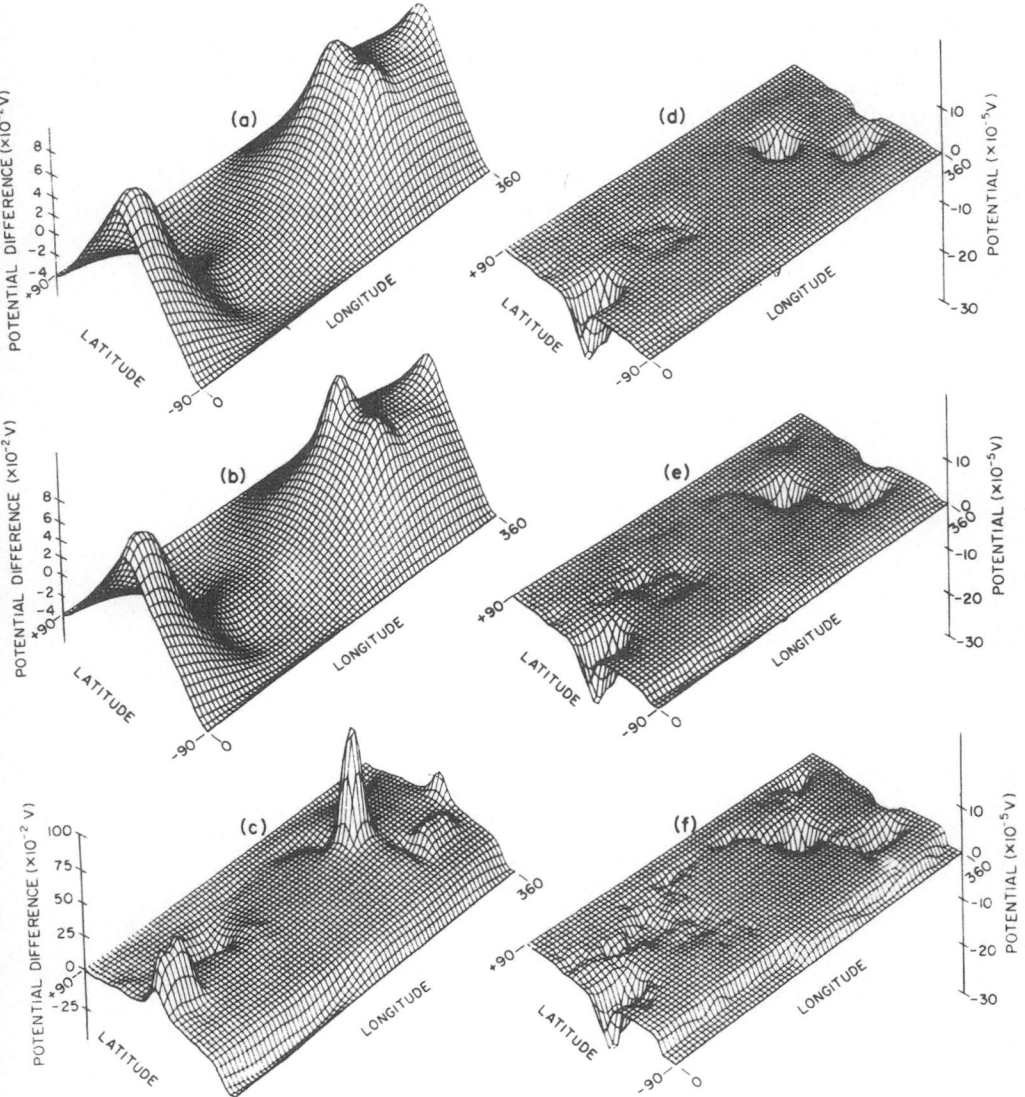

Fig. 5.6a – f. Perspective illustrations of calculated electric potential of the fair weather field along various constant conductivity surfaces. **a, b,** and **c** illustrate the calculated potential difference against the globally averaged atmospheric potential $\Phi_A = 291$ kV at, respectively, roughly 105 km, 50 km, and 25 km altitude. The illustrations **d, e,** and **f** are the potentials at about 8 km, 4 km, and 2 km, respectively. (Hays and Roble 1979)

flowing upward into the ionosphere. Panels d to f in Fig. 5.6 give the potential Φ below the thunderstorm heights (<8 km). The potential is seen to be negative below the thunderstorm areas corresponding to electric currents flowing upward. The potential is positive in the fair weather regions corresponding to downward-directed currents. The series of spherical functions in these calculations was developed up to terms of $n = 37$, giving an effective 5° grid. Clearly, fine structures of the field cannot be adequately determined by such a model.

6 Lightning

Lightning discharges are breakdown electric currents starting in regions where the electric field locally exceeds about 400 kV/m. Most lightning events occur in connection with thunderclouds. Lightning has been observed, however, during volcano eruptions and in dust storms. In this chapter, we outline the normal phenomenon of lightning. Ball and bead lightning are rare events and not well documented. We will not discuss these exotic forms of lightning. The reader is referred to Barry (1980).

6.1 Phenomenology of Lightning

Lightning in thunderclouds may be initiated by the emission of corona from the surface of rain drops, highly deformed by a strong electric field (Latham and Stromberg 1977). Ground discharges normally transport negative charge (electrons) from the lower part of the cloud to the ground and are therefore part of the global circuit (R_1 in Fig. 5.5). Cloud discharges (discharges within the cloud, between the clouds, and between cloud and air) are short circuits within the generator and redistribute the charge inside the clouds (Berger 1977; Ogawa 1982; Uman and Krider 1982).

Each ground discharge is generally made up of one or more intermittent partial discharges (Fig. 6.1). A total lightning discharge, the time duration of which is about 1/3 s, is called a flash. Each component discharge, lasting of the order of tens of milliseconds, is called a stroke. Each lightning stroke is preceded by a barely luminous predischarge, the so-called leader process, which produces a negatively charged ionized path between cloud and ground for the return stroke to follow. The first leader is called a stepped leader because it moves in steps of about 50 m downward and pauses between the steps. It takes about 20 ms to reach the ground from the cloud base near 3 km altitude. A negative charge of about 5 C (mainly electrons) is distributed along the leader channel during this process, and a current of the order of 100 A flows. The channel diameter is of the order of 1 cm. A corona sheath with a diameter of about 1 m envelops the highly conducting channel.

When the stepped leader has approached to within 5 to 50 m of the ground, a positively charged streamer from some point on the earth comes up to meet it, and then commences the return stroke which travels up the ionized

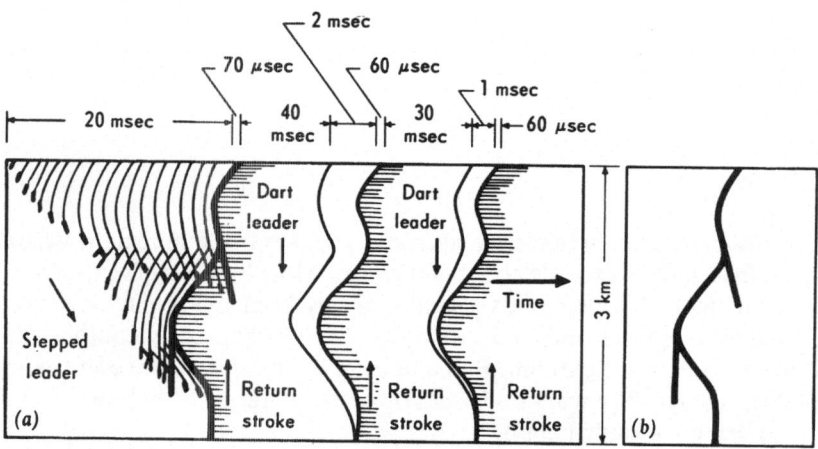

Fig. 6.1. Schematical illustration of time sequence of luminous features of a lightning flash. For clarity, the time scale has been distorted. The *right panel* shows the same flash as recorded with a camera with stationary film. (Uman 1969)

channel established by the leader. The negative charge in the channel is lowered to the ground. The return stroke can be observed from optical studies as an upward movement of a luminous wave front with a velocity that reaches 0.1 to 0.3 of the speed of light. This wave front arrives at the cloud base in about 70 μs. An upward-directed positive electric current flows during this event (the return stroke current) which reaches peak amplitudes of 10 to 100 kA within a few tens of microseconds. Figure 6.2 shows examples of return stroke currents measured at a lightning rod on the top of a tower.

The visible part of the return stroke channel is more or less vertical with some sideward branches. The upper part of the channel hidden in the cloud often has a horizontal extention which exceeds its vertical extention (MacGorman et al. 1981).

After the first return stroke, a next stroke may start from a higher charge center initiated by a dart leader. This dart leader reestablishes the ionization in the original channel and propagates downward faster than the stepped leader and without steps. The subsequent return stroke following the dart leader is, in general, not branched and reaches its maximum current strength within a few microseconds. On the average, two or three return strokes occur during one flash. However, as many as 26 strokes have been recorded during a single flash (e.g., Salanave 1980). A continuous current of the order of 100 A often flows from the ground to the negatively charged portion of the cloud in the intervals between subsequent return strokes.

Less common types of ground discharges are upward-moving positive leaders followed by negative downward-moving return strokes, mainly initiated from tall towers (Berger 1977), and downward-moving positive leaders followed by upward-moving negative return strokes, mainly occurring in

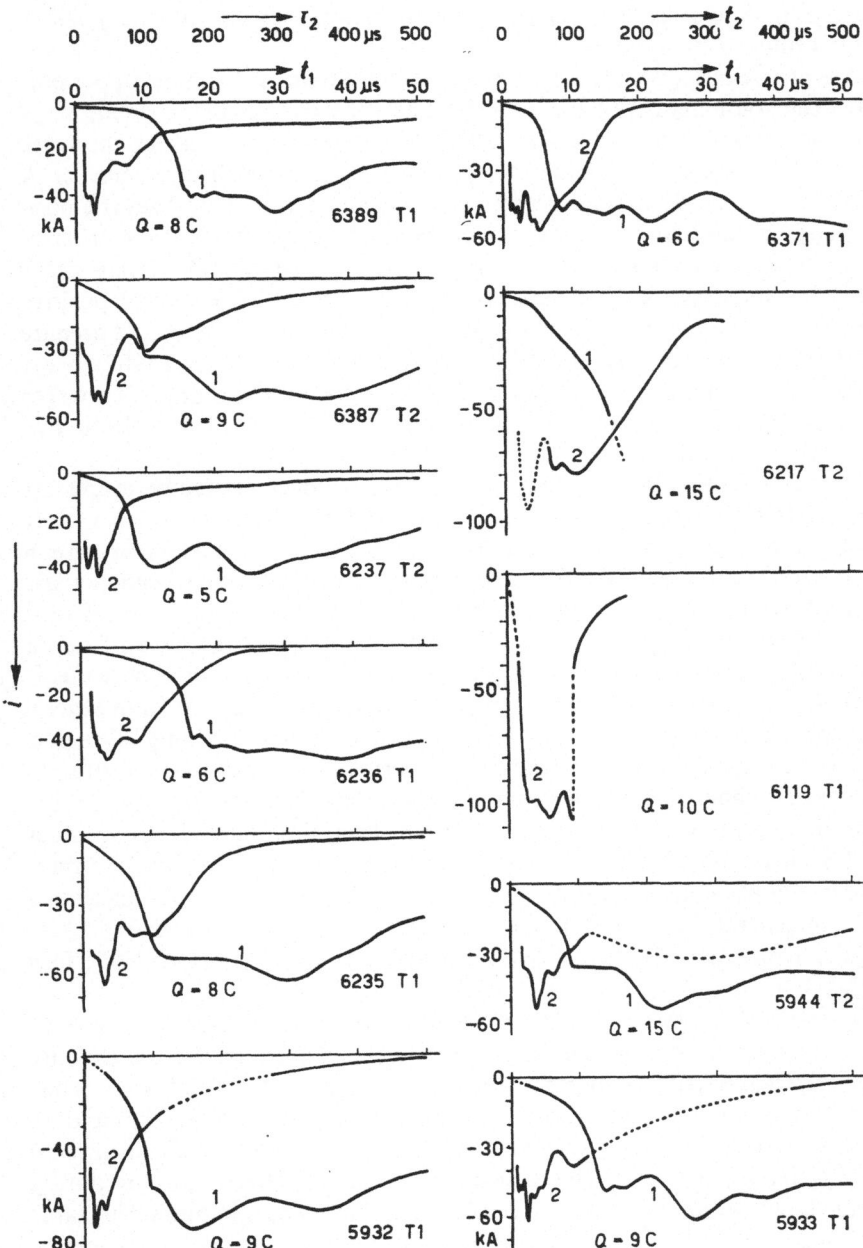

Fig. 6.2. Examples of lightning return currents recorded at a tower in San Salvatore, Switzerland. Note the two different time scales. The direction of the current is defined negative if the current flows upward. (Coordinate system of Fig. 14.1b). (Berger 1977)

winter thunderclouds with large horizontal displacement of the positive charge center (Takeuti et al. 1978).

A typical cloud discharge, initiated by a slow streamer from the upper portion of the boundary between positive and negative charges in the cloud, propagates downward as well as horizontally. When the streamer reaches the space charge concentration of opposite (negative) sign, a recoil streamer, or K stroke, runs back along the already ionized channel and neutralizes the positive charge in the channel. Maximum currents of K strokes are of the order of 1 kA. The K stroke can repeat several times. Streamers and K strokes occur during the intervals between successive return strokes. They lower positive charge to the lower part of the cloud (transport electrons upward) and prepare the stage for subsequent strokes (Brook and Ogawa 1977; Ogawa 1982). Most lightning events are cloud discharges. The ratio of cloud to ground discharges increases from a factor of 3 at midlatitudes to a factor of 5 to 10 at low latitudes (Prentice 1977).

Luminous events in lightning channels start to occur at temperatures exceeding 10000 K. The visible spectrum of the lightning stroke consists mainly of neutral nitrogen and oxygen emissions and the H_α line. The temperature in return strokes can reach 30000 K, and the pressure in the channel can increase to a peak of more than 10^6 Pa (Orville 1977). A pressure shock wave is thus generated which propagates away from the channel with a speed of about 3 km/s. This speed decreases rapidly as the shock front expands. The acoustic signal of the shock front is heard as thunder. Acoustic signals from several small segments of a tortuous channel interfere to produce "rolling" thunder. A signal arising from a direction nearly perpendicular to a larger straight segment of a channel can be heard as a thunderclap. Thunder travels with the speed of sound ($\simeq$ 330 m/s). The maximum spectral amplitude of a thunder signal is near 100 Hz. The atmosphere attenuates, scatters, and refracts thunder signals depending on the prevailing temperature, turbulence and winds. In general, thunder from ground flashes cannot be heard beyond 5 km. Thunder from cloud flashes is seldom heard beyond 25 km (Hill 1977; Few 1982).

Optical instruments on orbiting satellites can measure the visible and infrared emission of strong lightning events. The satellites of the US Defense Meteorological Satellite Program have circular sun synchronized polar orbits around the earth with periods of 101.56 min at an altitude of 830 km and an inclination of 98.7°. Orbital precession keeps these satellites fixed in local time throughout the year. The instrument is able to detect lightning during daytime hours within an area of about 1.5×10^6 km^2. The minimum detectable signal is 5×10^{-4} W/m^2, corresponding to an unattenuated source power of 4×10^9 W. Lightning flashes with peak optical power in excess of 3×10^{12} W (superbolts) have sometimes been observed. Since only the strongest sources can be detected, the detection efficiency of the instrument is about 2%.

Figure 6.3 shows the lightning distribution for the period September 10 through October 11, 1977 during dawn (a) and dusk (b). More lightning activ-

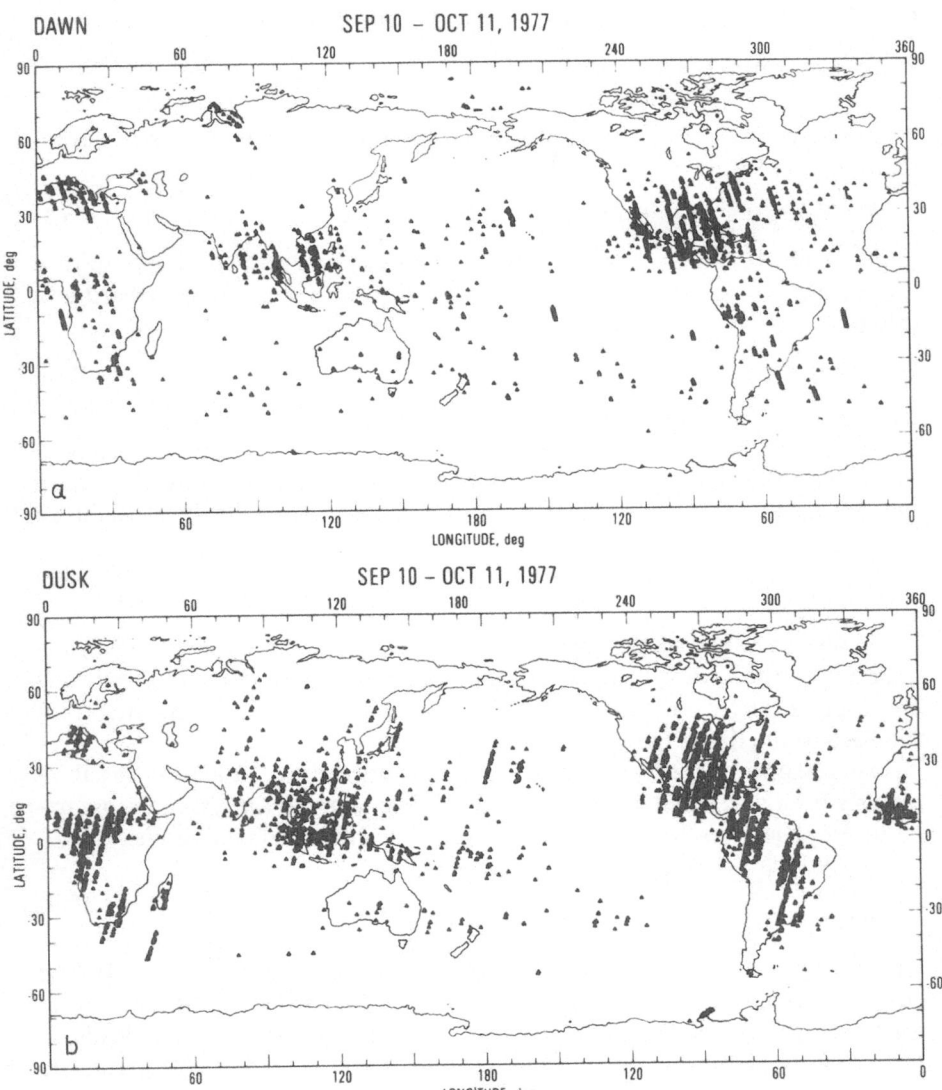

Fig. 6.3. Lightning distribution during dawn and dusk for the period September 10 to October 11, 1977, determined from satellite observations. (Orville 1982)

ity is observed at dusk than at dawn over the continents. Furthermore, lightning activity over the oceans is considerably less than over the continents. The absolute mean flash rate is estimated to be of the order of 100 flashes per second. The flash rate increases from midlatitudes (40°) to the equator by a factor of about 10. The ratio of land to ocean flashes is also about 10 (Orville 1982).

6.2 Bruce-Golde and Transmission Line Models
of Return Stroke Currents

The observations of return stroke currents (Fig. 6.2) suggest a time dependence of the form

$$J = J_0[\exp(-\alpha t) - \exp(-\beta t)] \quad (t \geqq 0) \tag{6.1}$$

(e.g., Golde 1977). On the other hand, optical observations of the upward
propagating luminous wave front show that the stroke expands upward at a
rate of

$$w = w_t \exp(-\eta t), \tag{6.2}$$

where typical values of α, β, J_0, w_t, and η are given in Table 6.1.

Bruce and Golde assumed that the luminosity directly reflects the behavior
of the electric currents. The electric current is therefore taken as uniform in
the channel up to the tip of the wave front and zero above the tip at height h_t:

$$\begin{aligned} J(z, t) &= J(0, t) \\ J(z, t) &= 0 \end{aligned} \quad \text{for} \quad \begin{aligned} z &\leqq h_t \\ z &> h_t. \end{aligned} \tag{6.3}$$

While the return stroke current propagates upward, eventually decreasing
in amplitude according to Eq. (6.1), the charge originally stored in the corona
envelope of the leader is lowered to the ground along the entire channel from
the base to the tip of the wave front.

This height-independent charge transfer is unrealistic. Models have subse-
quently been developed where the current wave form on the ground is assumed to propagate up the channel like the current in a transmission line:

$$\begin{aligned} J(z, t) &= J(t - w/c) \\ J(z, t) &= 0 \end{aligned} \quad \text{for} \quad \begin{aligned} z &\leqq h_t \\ z &> h_t, \end{aligned} \tag{6.4}$$

with c the speed of light (Leise and Taylor 1977; Dennis and Pierce 1964). The
current terminates at the top of the channel, and no reflection of the wave at
the top is included in the model. Clearly, this is again unrealistic and leads, in
fact, to artificial abrupt field changes ("mirror images") if these currents are
applied for the calculation of electromagnetic radiation energy (Uman et al.
1975). A more refined model is due to Price and Pierce (1977).

Table 6.1. Parameters of Bruce-Golde's formula of typical lightning return stroke currents.
(Adapted from Dennis and Pierce 1964)

	J_0 (kA)	α (s^{-1})	β (s^{-1})	w_t (m/s)	η (s^{-1})
First return stroke	30	2.0×10^4	2.0×10^5	8×10^7	3×10^4
Subsequent return stroke	10	1.4×10^4	6.0×10^6	1×10^8	0

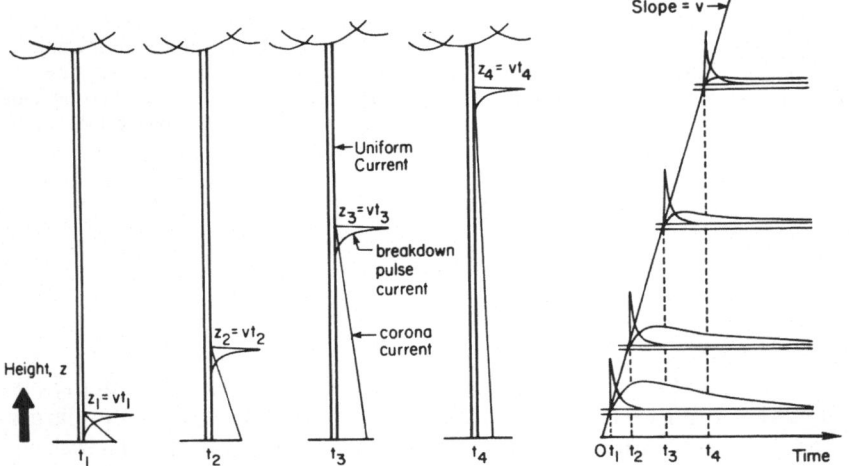

Fig. 6.4. Return stroke model of Lin et al. (1980) consisting of three components: a short duration pulse propagating upward, a uniform continuous current, and a corona current

A model which tries to combine the Bruce-Golde and the transmission line model is due to Lin et al. (1980). It considers three separate current components: (1) a short duration upward-propagating pulse associated with the electric breakdown of the upward-propagating return stroke wave front. It traverses the channel with the velocity of the return stroke wave front and is treated by the transmission line theory according to Eq. (6.4) (Fig. 6.4); (2) a uniform current that may already be flowing or may start to flow soon after the return stroke begins; and (3) a coronal current which is caused by the radially inward and then downward movement of the charge initially stored in the corona sheath around the leader channel. The coronal current is idealized by a number of current sources distributed along the channel. Each source is turned on when the peak of the breakdown pulse (1) reaches the altitude of the source.

More than ten free parameters are available in this model to adapt it to observed wave forms of currents and fields from lightning return strokes. Since the breakdown current becomes unimportant after about 10 μm, the "mirror image" (which is still present in the model) does not greatly influence the total field structure.

6.3 Resonant Wave Guide Model of Return Strokes and K Strokes

The transmission line models discussed in the previous section do not properly take into account the finite length of the channel. The boundary conditions

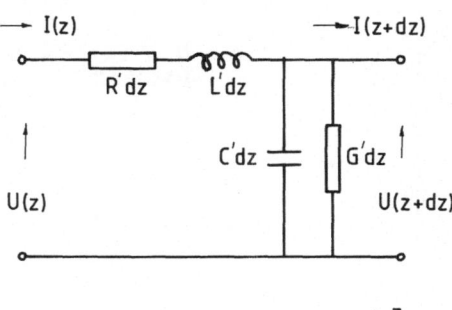

Fig. 6.5. Element of a transmission line with resistance per unit length R', inductance per unit length L', capacitance per unit length C', and conductance per unit length G'. U and J are voltage and current

are important, however, since the spectral wavelengths inferred from the observed wave forms of return strokes are of the order of 40 km, which is larger than the channel lengths. One might expect, therefore, that the channel behaves like a resonant wave guide.

In order to apply wave guide theory to the lightning phenomenon, we assume that the lightning channel is a straight homogeneous transmission line of length h. A transmission line section as shown in Fig. 6.5 consists of series resistance per unit length $R' = R/h$, series inductance $L' = L/h$, shunt conductance $G' = G/h$, and shunt capacitance $C' = C/h$. The relationship of current and voltage between the height z and the base at $z = 0$ is (e.g., Magid 1972; Oetzel 1968)

$$J(z, t) = [J_0 \cos (Kz/h) + (iU_0/Z) \sin (Kz/h)] \exp (-i\omega t)$$
$$U(z, t) = [iJ_0 Z \sin (Kz/h) + U_0 \cos (Kz/h)] \exp (-i\omega t) \tag{6.5}$$

with

$$iK = [(R - i\omega L)(G - i\omega C)]^{1/2}; \quad Z = [(R - i\omega L)/(G - i\omega C)]^{1/2} .$$

The boundary conditions for a return stroke are

$$J(h, t) = 0; \quad U(0, t) = 0 \quad \text{for} \quad t \geqq 0; \quad J(z, t) = 0 \quad \text{for} \quad t < 0 \tag{6.6}$$

because the conducting channel ends at $z = h$, and the conductivity is much smaller outside the channel than inside. According to Uman (1969), the channel conductivity is of the order of $\sigma = 10^4$ S/m. The channel is in contact with the well-conducting earth at $t > 0$ so that the channel base becomes a short circuit.

For our purpose, the shunt conductance G is negligible ($G = 0$). Inserting Eq. (6.6) into (6.5), it follows that the normalized vertical wave number K is determined by

$$K = K_n = (2n - 1) \pi/2 \quad (n = 1, 2, 3 \ldots), \tag{6.7}$$

while the frequency becomes the eigenfrequency of the n-th mode given by

$$i\omega = i\omega_n = \begin{Bmatrix} \alpha_n \\ \beta_n \end{Bmatrix} = R/(2L) \mp [R^2/(4L^2) - K_n^2/(LC)]^{1/2}, \tag{6.8}$$

indicating that only individual standing waves of eigenfrequency ω_n and vertical wavelength of

$$\lambda_n = 2\pi h/K_n = 4h/(2n-1) \tag{6.9}$$

can be excited in the channel. Their current and voltage amplitudes are now

$$J_n(z, t) = \bar{Q}_n \cos(K_n z/h) dL_n(t)/dt$$
$$U_n(z, t) = -\bar{Q}_n R \alpha_n \beta_n/[K_n(\beta_n + \alpha_n)] \sin(K_n z/h) L_n(t) , \tag{6.10}$$

and the electric charge density can be determined from the continuity equation of the electric current [Eq. (14.18)] as

$$q_n(z, t) = \bar{Q}_n K_n/V \sin(K_n z/h) L_n(t) , \tag{6.11}$$

with

$$L_n(t) = [\beta_n \exp(-\alpha_n t) - \alpha_n \exp(-\beta_n t)]/(\beta_n - \alpha_n) \tag{6.12}$$

the time function of q_n and U_n, and

$$\bar{Q}_n = F \int_0^h q_n(z, 0) \, dz \tag{6.13}$$

the total electric charge stored in the channel at time $t = 0$. $V = hF$ is the volume of the channel, and F is its cross section.

The total dissipated energy within the channel is

$$W_{\text{dis}} = (1/h) \int_0^\infty \int_0^h R J_n^2 dt \, dz = R \bar{Q}_n^2 \alpha_n \beta_n/[4(\alpha_n + \beta_n)] . \tag{6.14}$$

The form of $J_n(z, t)$ in Eq. (6.10) is identical with the Bruce-Golde formula (6.1) at $z = 0$, if (α_n, β_n) are real and if the current amplitude in Eq. (6.1) is replaced by

$$J_0 = -\bar{Q}_n \alpha_n \beta_n/(\beta_n - \alpha_n) . \tag{6.15}$$

(α_n, β_n) become complex if $K_n^2 > R^2 C/(4L)$, and the function $L_n(t)$ in Eq. (6.12) must be replaced by

$$L_n(t) = (\gamma_n \sin \delta_n t/\delta_n + \cos \delta_n t) \exp(-\gamma_n t) , \tag{6.16}$$

with

$$\gamma_n = (\beta_n + \alpha_n)/2; \quad i\delta_n = (\beta_n - \alpha_n)/2; \quad (\gamma_n, \delta_n) \quad \text{real}.$$

We call the aperiodic wave form of Eq. (6.12) $[(\alpha_n, \beta_n)$ real] a type 1 wave and the damped oscillating wave form (6.16) $[(\gamma_n, \delta_n)$ real] a type 2 wave. The transition from type 1 to type 2 occurs at $\delta_n = 0$, where one has

$$J_n(z, t) = -\bar{Q}_n \gamma_n^2 \cos(K_n z/h) t \exp(-\gamma_n t)$$
$$U_n(z, t) = -\bar{Q}_n R \gamma_n/(2K_n) \sin(K_n z/h)(1 + \gamma_n t) \exp(-\gamma_n t) . \tag{6.17}$$

A more sophisticated calculation of electromagnetic wave propagation in a cylindrical straight wire of diameter d, length h, and electric conductivity σ re-

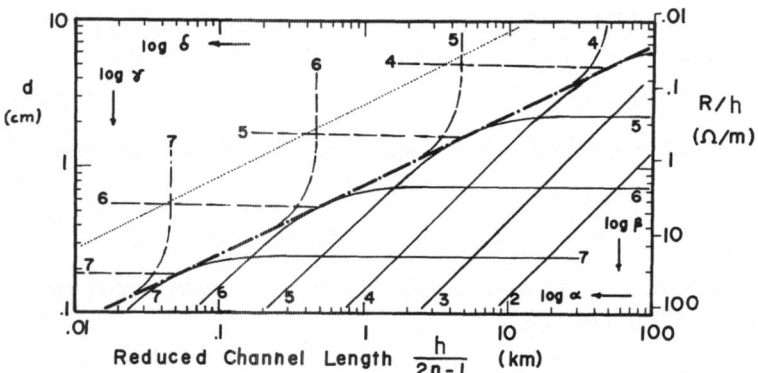

Fig. 6.6. Diagram connecting the channel diameter d (or resistance per unit length R/h) with the reduced channel length $h/(2n-1)$ in a cylindrical model of lightning currents. n is the mode number. $n = 1$ corresponds to the dominant mode. The parameters (α, β) and (γ, δ), respectively, belong to the type 1 and type 2 wave forms. The *thick dash-dotted line* separates the regimes of type 1 and type 2 wave forms (Volland 1982). The model is not valid in the *upper left beyond the dotted line*. The channel conductivity is assumed to be $\sigma = 10^4$ S/m.

veals that the transmission line elements in Fig. 6.5 are related to the wire elements d, h, and σ as follows (Volland 1982):

$$R = 4h/(\pi\sigma d^2); \quad L_n = R/(\alpha_n + \beta_n) = R/(2\gamma_n); \quad C_n = h^2/(L_n c^2). \quad (6.18)$$

c is the speed of light. Figure 6.6 shows the eigenvalues (α_n, β_n) and (γ_n, δ_n), respectively, as functions of $d(\sigma)^{1/2}$ (left ordinate), or R/h (right ordinate) and of the reduced channel length $h_n = h/(2n-1)$. The transition from the (α, β) regime to the (γ, δ) regime occurs at the thick dash-dotted line in Fig. 6.6 where $\delta = 0$. That transition takes place at a critical resistance

$$R_{\text{cri}} \simeq 3.5 - 0.225 \log \gamma; \quad (\gamma \text{ in s}^{-1}). \tag{6.19}$$

For a typical value of $\sigma d^2 = 1$ ($d = 1$ cm for $\sigma = 10^4$ S/m), the critical resistance becomes $R_{\text{cri}} \simeq 2.3$ kΩ. Type 1 wave forms are, therefore, expected to occur if the channel resistance is larger than R_{cri}. Type 2 wave forms occur at values $R < R_{\text{cri}}$.

Figure 6.7a shows the vertical structure of current and charge density from Eqs. (6.10) and (6.11) for wave numbers $n = 1$ and 2, assuming $\bar{Q}_n < 0$: In the case of the first mode ($n = 1$), the vertical current is upward (positive) within the whole channel, and the total negative charge is drained to the earth. The vertical wavelength of this mode is $\lambda_1 = 4h$. For the second mode ($n = 2$), only the negative excess charge within the lower part of the channel ($z < h/3$) is drained to the earth. It corresponds to an upward current in this part of the channel. In the upper part, an internal downward current discharges the two regions of opposite polarity. Therefore, the effectiveness of the higher order modes to transport charge out of the channel decreases as $1/(2n-1)$ with increasing wave number.

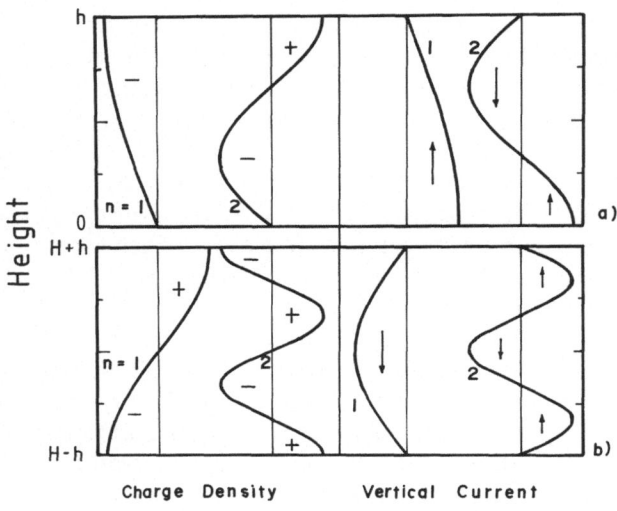

Fig. 6.7a, b. Height distribution of charge density q (*left*) and vertical electric current J (*right*) for the two modes of order $n = 1$ and 2. **a** Return stroke model; **b** intracloud K stroke model. In model **a**, the total charge $\bar{Q}_n$ is negative. In model **b**, the total charge is zero, and is positive in the *upper half* of the channel

Charge Density   Vertical Current

This model can also be applied to intracloud K strokes. If a conducting channel exists for a sufficiently long time without contact to the earth, resonant modes may occur in that channel which must now obey the boundary conditions $J = 0$ at both ends of the channel. If we center the intracloud channel at a height H, the channel has a length of $2h$ and reaches from $H - h \leq z \leq H + h$. Its current and charge configuration in the upper half of the channel is the same as for the return stroke (Fig. 6.7b). The total charge stored in the upper half of the channel is given by Eq. (6.13). That charge is neutralized in the channel during the stroke.

6.4 Excitation of Resonant Modes

The real charge distribution in a return stroke channel before breakdown may be rather complex. It can be Fourier decomposed at time $t = 0$ by

$$\bar{Q}_n = 2V/(K_n h) \int_0^h q(z, 0) \sin(K_n z/h)\, dz \tag{6.20}$$

Each term $\bar{Q}_n$ excites its corresponding mode independently. The superposition of all modes having different temporal and height structures gives rise to charge transport with finite velocity in the same manner as in any other resonance system (e.g., the transport of a mechanical pulse on a piano string).

In a real inhomogeneous and tortuous channel, resonance loses its meaning for the higher order modes which have wavelengths that are small compared with the channel length. This is certainly the case for $n \gtrsim 10$ or for wavelengths $\lambda_n \lesssim h/5$. These wave modes propagate nearly as free waves. They become partially reflected at inhomogeneities and are attenuated along their propagation paths. They are generated at all times during a flash. However, they are ineffective for transport of electric charge because they interfere destructively. Their contribution to channel heating is probably of minor importance. These modes generate the quasi-continuous radio noise at frequencies greater than about 100 kHz (see Sect. 7).

Waves with intermediate wave numbers ($3 \lesssim n \lesssim 10$) have wavelengths comparable with the large-scale inhomogeneities of the channel. However, their wavelengths are too large to allow propagation as free waves. Therefore, these waves cannot develop to their full pulse forms, and mode coupling into waves of other wave numbers is expected.

The waves of lowest wave number are the only modes where resonance can lead to a full development of their wave forms. In particular, the first mode ($n = 1$), having a wavelength of four times the channel length, is least influenced by the real channel configuration and most effectively transports and redistributes electric charge. This mode also heats the channel to temperatures where luminous events can be observed. A discharging process where the first mode is involved probably starts after the channel has established a quasi-stable configuration.

The wave modes are standing waves excited simultaneously after breakdown along the whole channel. This is in apparent disagreement with the observational fact that a luminous wave front propagates upward with a velocity of a fraction of the speed of light. One has to bear in mind, however, that luminosity and the electric current are two distinct phenomena. The channel is heated by Joule heating due to the dissipation of the electric current. Luminous events do not occur before the channel has reached a temperature of about 10000 K (Orville 1977). Since the current amplitude of the dominant first mode has its maximum on the ground, the point where the critical temperature of luminosity is reached "travels" from the bottom of the channel upward.

We can estimate the heating process starting from the first law of thermodynamics (neglecting heat losses by conduction and/or radiation) which gives

$$V \varrho c_p dT/dt \simeq R J^2, \tag{6.21}$$

with ϱ the gas density, c_p the specific heat at constant pressure, T the temperature, R the resistance of the channel, V its volume, and $J \simeq J_1$ the stroke current simplified by the current of the first mode. Since J_1 decreases with height according to a cosine law [see Eq. (6.10)], the height where the threshold temperature T_{cr} for luminous events is reached is a function of time t_{cr}:

$$z(t_{tr}) \simeq (2h/\pi) \arccos [w_{\text{therm}}/w_{\text{joule}}]^{1/2}, \tag{6.22}$$

with

$$w_{\text{therm}} \simeq \varrho c_p (T_{cr} - T_0) \simeq 10 \text{ MJ/m}^3 \tag{6.23}$$

the thermal energy per volume to heat the channel to the temperature $T_{cr} \simeq 10000$ K, T_0 the temperature of the channel prior to the stroke and

Table 6.2. Parameters of lightning strokes. Parameters (α, β) and (γ, δ), respectively, current amplitude $J_0 = -\alpha\beta\bar{Q}/(\beta - \alpha)$, and maximum charge $\bar{Q}$ of observed lightning currents and sferic wave forms, and derived channel parameters: channel diameter d, resistance R, inductance L, capacitance C, maximum voltage $\bar{U}$, magnetic moment $\bar{M}$, total dissipated energy W_{dis}, ratio between radiated and dissipated energy W_{rad}/W_{dis}, and maximum spectral frequency f_{max}. The cloud-to-ground (G) first return stroke (R) of type 1 $(G-R_1)$ is derived from the Bruce-Golde parameters in Table 6.1. The type 1 return stroke $(G-R_2)$ simulates observations of Krider et al. (1977) (reproduced in Fig. 7.4). The type 2 first return stroke $(G-R_3)$ is that of Fig. 7.2 at 200 km distance. The intracloud (C) K discharge $(C-K)$ of type 2 simulates observations of Weidman and Krider (1978) (Fig. 7.3)

Type	Stroke	α (μs^{-1}) / γ (μs^{-1})	β (μs^{-1}) / δ (μs^{-1})	J_0 (kA)	d (cm)	h (km)	R (kΩ)	L (mH)	C (pF)	$\bar{Q}$ (C)	M (C km)	$\bar{U}$ (MV)	W_{dis} (MJ)	W_{rad}/W_{dis} (%)	f_{max} (kHz)
1	$G-R_1$	0.02	0.2	30	1.55	7.89	4.20	19.3	35.9	−1.35	−13.6	−66.3	69.7	2.1	10.1
	$G-R_2$	0.1	0.5	4	0.96	2.22	3.08	5.2	10.5	−0.03	−0.09	−5.3	0.13	2.9	35.6
2	$G-R_3$	0.012	0.024	7	4.46	18.0	1.15	50.7	70.6	−0.47	−10.7	−10.8	3.75	7.3	4.27
	$C-K$	0.1	0.2	−1	1.61	2.16	2.11	11.3	18.4	0.008	0.044	2.87	0.017	15.9	35.6

$$w_{\text{joule}} \simeq (RJ_0^2/V) \int_0^{t_{cr}} [\exp(-\alpha_1 t) - \exp(-\beta_1 t)]^2 dt \leq 2W_{\text{dis}}/V \qquad (6.24)$$

the maximum Joule heating per volume at the base of the channel at time t_{cr}, J_0 the current amplitude from Eq. (6.15), and W_{dis} the total dissipated energy in the channel from Eq. (6.14). Evidently, w_{therm} must be smaller than $2W_{\text{dis}}/V$ in order that luminous events can be observed at all. The degree of luminosity depends mainly on the ratio between charge and cross section of the channel. Of course, this estimate is very rough. One must also consider heat losses to the environment and temperature effects on the plasma parameters (Hill 1963). Moreover, the temperature T_0 of the leader channel is not well known.

Table 6.2 gives some values for W_{dis}, estimated from lightning data and from the theory outlined in this section. Type 1 return strokes give probably the brightest optical signals. Type 2 strokes are, in general, not so bright because of the larger cross sections of their channels.

7 Sferics

Lightning channels behave like huge antenna systems which radiate electromagnetic energy. The signals are of impulsive nature at frequencies below about 100 kHz. They gradually evolve into red noise at higher frequencies (Fig. 7.1). The pulses arise from coherent electric currents in lightning channels during return strokes or K strokes. They are called atmospherics or simply "sferics" for short. Sferics are useful tools for locating thunderstorms over greater distances. The high frequency noise spectrum is discussed by Lewis (1982) and Spaulding (1982).

7.1 Observed Wave Forms

The observed wave forms of sferics depend on the antenna characteristics of the lightning channel, the dispersion characteristics of the wave guide between earth's surface and ionosphere, and the transmission function of the receiver. Since about 100 flashes per second are excited all over the world, interference between two or more sferics may lead to complex wave forms. In order to study individual wave forms, one generally selects sufficiently large isolated pulses arising from near sources or from strong sources at greater distances.

Since each receiver has a finite bandwidth and a limited time constant, it is impossible to measure the whole spectrum of the electromagnetic field of lightning events with a single instrument. A receiver with a large time constant measures mainly slow changes of the electric field of nearby flashes. Figure 7.1 shows a typical sequence of electric field changes due to several flashes. The return stroke events are preceded by leader processes. K strokes occur between successive return strokes, and a slow increase of the electric field following each return stroke may be attributed to a continuous current.

The vertical component of the electric field plotted in Fig. 7.1 is defined as positive downward. Furthermore, the field is defined as zero at the beginning of a flash (or of a stroke in Figs. 7.1 to 7.4). We use a different definition throughout this book. The electric field is positive in the upward direction (Fig. 14.1a), and the field disappears with disappearing charge. Thus, a transformation

$$E_z'(t) = E_z(0) - E_z(t) \quad \text{for} \quad t \geq 0 \tag{7.1}$$

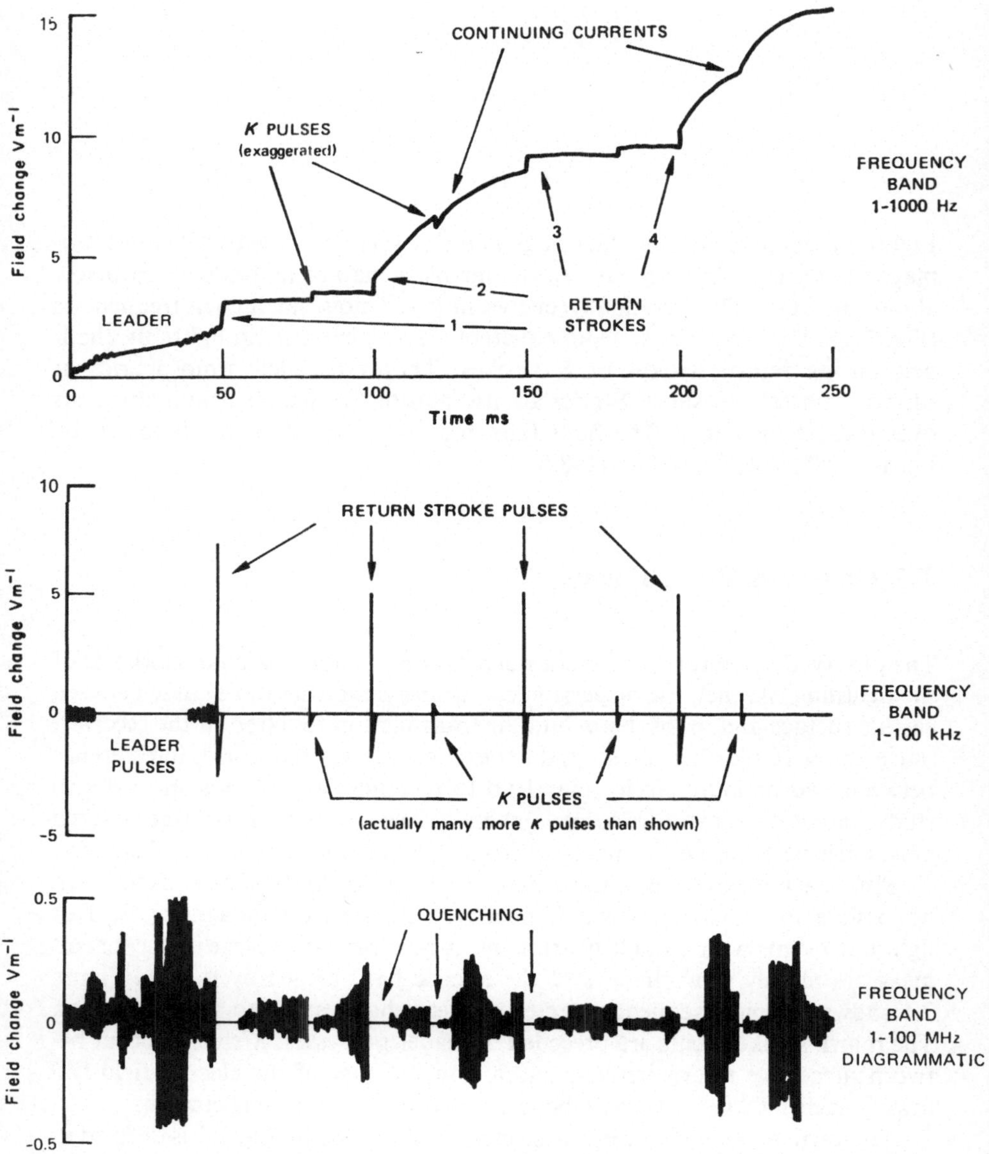

Fig. 7.1. Electric field changes due to a series of flashes observed at about 50 km distance in three different frequency bands (Pierce 1977). Orientation of E is positive downward (Fig. 14.1b)

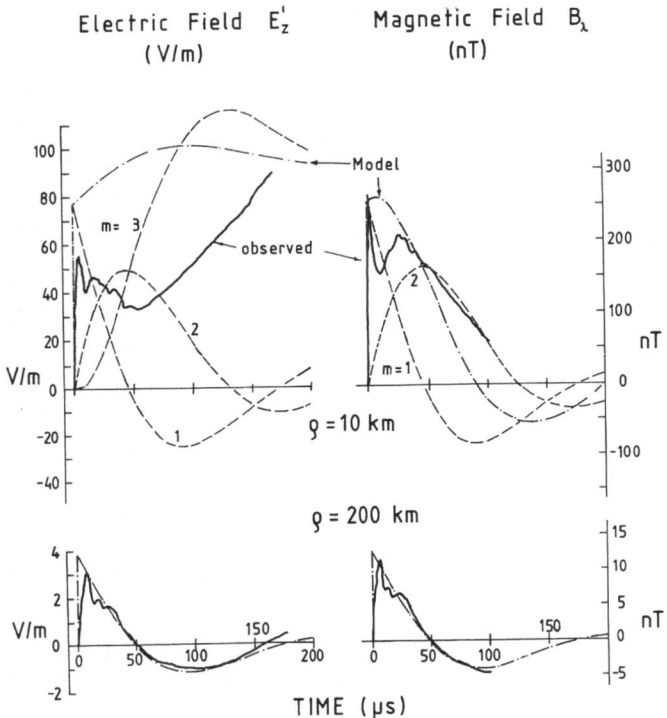

Electric Field $E_z^!$
(V/m)

Magnetic Field B_λ
(nT)

Fig. 7.2. Typical wave forms of vertical electric field (*left*), and azimuthal magnetic field (*right*) of first return strokes at 10 km and 200 km distance from the origin (*solid curves*) (Lin et al. 1979), and model calculations of dipole fields from Eqs. (7.6) and (7.7) with the numbers of (γ, δ, J_0) from R_3 in Table 6.2. *Dashed lines* give separately, the radiation components ($m = 1$), the induction components ($m = 2$), and the electrostatic component ($m = 3$). *Dash-dotted lines* give the total field of the dipole. Note that $E_z^!$ according to Eq. (7.1) is plotted

is necessary where $E_z^!$ is the electric field in Figs. 7.1 to 7.4, and E_z is the field according to our definition.

Instruments with higher time resolution can measure the wave forms of return strokes, leaders and K strokes. Figure 7.2 (solid lines) shows typical electric and magnetic fields of return strokes observed at distances of 10 and 200 km from their origin. Figure 7.3 shows electric fields of K strokes at a distance of about 25 km. The typical far field behavior of the wave forms in Figs. 7.2 and 7.3 is their bipolar character with a reversal from positive to negative fields after about 30 to 50 μs in the case of the return strokes at distances greater than 50 km, and after 5 to 10 μs in the case of K strokes. Figure 7.4 gives an example of wave forms of dart-stepped leaders from about 50 km distances preceding return strokes of monopole character (type 1).

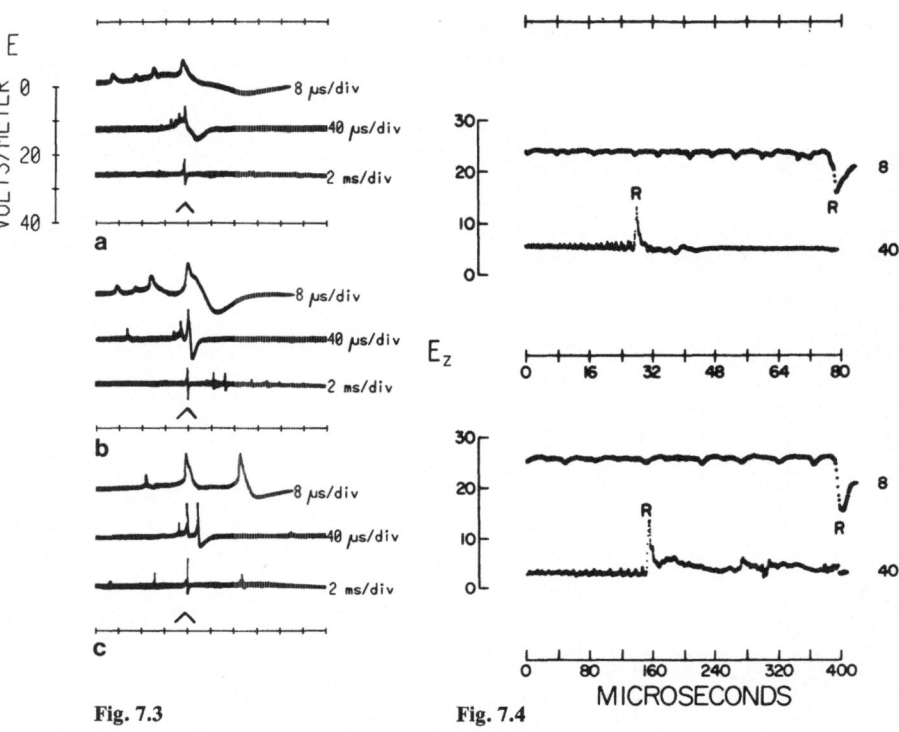

Fig. 7.3   **Fig. 7.4**

Fig. 7.3 a – c. Wave forms of electric fields from K strokes at distances of 15 to 30 km. Each discharge is shown on three time scales of 2 ms/div, 40 μs/div, and 8 μs/div (Weidman and Krider 1978). Orientation of electric field is positive upward (Fig. 14.1a)

Fig. 7.4. Wave forms of electric fields of dart-stepped leaders and return strokes (R) recorded in Arizona from lightning at about 50 km distance. Each discharge is shown in two time scales. Orientation of E in the slow time scale of 40 μs/div (*lower trace*) is positive downward. The fast time scale (8 μs/div) in the *upper trace* is inverted in orientation and magnified by a factor of two (Krider et al. 1977)

7.2 Electromagnetic Radiation from Lightning Channels

We want to determine in this section the electromagnetic field generated by lightning currents. We approximate the lightning channel by a vertical straight thin wire of length $(h+s)$ centered at height H in which a current $J(z, t)$ flows. It is $s = h$; $H > h > 0$ for K strokes, and $s = H = 0$ for return strokes. The electromagnetic field of such a line current in free space can be derived from the vector potential $A(0, 0, A_z)$ which has one nonvanishing component in the vertical direction given by (e.g., Sommerfeld 1952)

$$A_z = -\mu/(4\pi) \int_{H-s}^{H+h} \int_{t_r}^{\infty} J(z', t')/r \, dt' dz' \tag{7.2}$$

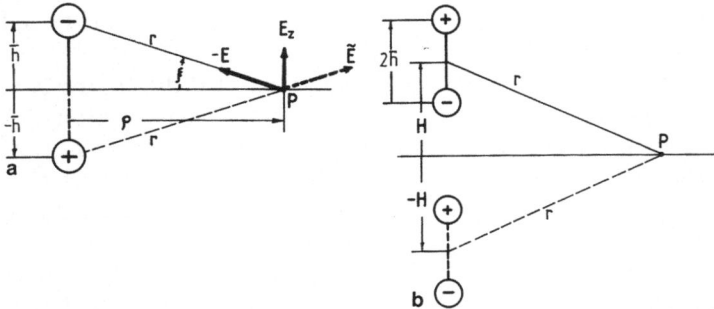

Fig. 7.5a, b. Geometry of lightning channels and their images. **a** Cloud-to-ground return stroke; **b** intracloud K stroke

$r = [\varrho^2 + (z - z')^2]^{1/2}$ is the distance between a channel element at $(0, z')$ and the observer at (ϱ, z) (in cylindrical coordinates; see Fig. 7.5), $t_r = t' - r/c$ is the retarded time. The time t' is the channel time. At $t' = 0$, the current starts to flow. The time $t = t' - r_0/c$ is the receiver time. At $t = 0$, the first signal reaches the receiver. r_0 is the shortest distance between channel and receiver. $\mu = 1/(\varepsilon c^2) = 4\pi \times 10^{-7}$ H/m is the permeability of free space.

The electric and magnetic field components are then derived from

$$E = -\partial^2 A/\partial t^2 + c^2 \nabla \nabla \cdot A$$
$$B = (\partial/\partial t) \nabla \times A ,$$

(7.3)

where, by definition, $E = 0$ at $t = \infty$.

At distances sufficiently far away from the origin $(r \gg h)$, r is only weakly dependent on z. If we furthermore assume a current configuration consisting of only the first resonant mode in Eq. (6.10) (dropping the index "1" for convenience), we arrive at the potential of a vertical electric dipole

$$A_z \simeq \mu \bar{Q} \bar{h}/(4\pi r) L(t) ,$$

(7.4)

where

$$\bar{h} = \int_{H-s}^{H+h} \cos[k(z - H)/h] dz = 2(h + s)/\pi$$

(7.5)

is an effective antenna length. Placing such a dipole over the highly conducting earth, one must add an image dipole at height $-H$, in order that the horizontal electric fields on the ground superpose to zero. An observer on the ground $(z = 0)$ then measures a vertical electric field given by

$$E_z \simeq -M/(4\pi\varepsilon)\{[L/r^3 + dL/dt/(cr^2)](1 - 3\cos^2\theta) + d^2L/dt^2/(c^2r)\sin^2\theta\}_{t=t_r},$$

(7.6)

and an azimuthal magnetic field given by

$$cB_\lambda \simeq M/(4\pi\varepsilon)[dL/dt/(cr^2) + d^2L/dt^2/(c^2r)]_{t=t_r}\sin\theta ,$$

(7.7)

with $\sin\theta = \varrho/r$; $\cos\theta = -H/r$, $M = 2\bar{Q}\bar{h}$ an electric dipole moment, and $L(t)$ from Eqs. (6.12) or (6.16).

The term with the $1/r^3$ dependence in Eq. (7.6) is called the electrostatic component because it describes quasi-electrostatic field changes due to redistribution of the electric charge. The terms with the $1/r^2$ dependence in Eqs. (7.6) and (7.7), called the induction component, describe the quasi-static electromagnetic field changes due to the electric channel currents. The terms with the $1/r$ dependence, finally, describe the far field radiation component.

The dashed lines in Fig. 7.2 are derived from Eqs. (7.6) and (7.7) based on the numerical values of a type 2 return stroke (R_3 in Table 6.2). Here, E_z' according to Eq. (7.1) is plotted. The dash-dotted curves represent the sum of the contributions from the three field components, indicating the relative importance of the radiation field, the induction field and the electrostatic field at various distances. The radiation field is an excellent approximation of the wave form at distances greater than about 50 km. The model also simulates the magnetic component reasonably well, even at near distances. The near-distance electric field is, however, not so well simulated by the dipole model. This is not surprising in view of the crude approximation.

Figure 7.6 shows calculated electric radiation fields based on the numerical values in Table 6.2. They simulate the radiation fields of the type 1 Bruce-Golde current, the type 1 R wave form from Fig. 7.4, and the two type 2 wave forms of the R stroke in Fig. 7.2 and the K stroke in Fig. 7.3. The channel parameters derived from these wave forms appear to be realistic and agree with data derived from other sources (Berger 1977; Hill 1979).

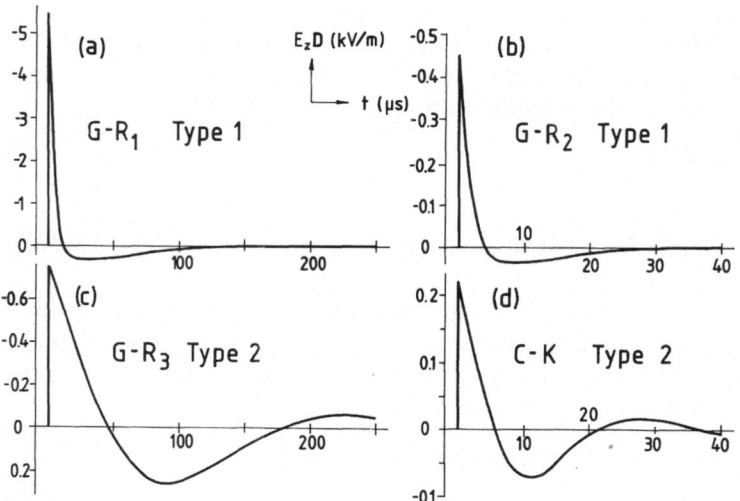

Fig. 7.6a – d. Electric radiation field component of radiation from a vertical electric dipole. The parameters (α, β) and (γ, δ), respectively, and of J_0 are taken from Table 6.2. The electric field is normalized to a distance of 1 km. Orientation of electric field of the return strokes is positive downward (Fig. 14.1b)

The total radiated energy of the first mode at large distances into the half-space above ground is

$$W_{rad} = (2\pi c/\mu) \int_0^\infty \int_0^{\pi/2} B_\phi^2 r^2 \sin\theta \, d\theta \, dt = \mu M^2 \alpha^2 \beta^2 / [24\pi c(\alpha+\beta)] \; . \quad (7.8)$$

Table 6.2 contains the ratio of W_{rad}/W_{dis}, which is a measure of the radiation efficiency of the lightning channel. Return strokes have efficiencies smaller than 10%, while K strokes may reach efficiencies of the order of 10% or more. More detailed calculations have been presented by Lin et al. (1980), Volland (1982), and Pathak et al. (1982).

7.3 Propagation Effects on Sferic Wave Forms

In the previous section, we have considered the ground as a perfectly conducting plane on which electromagnetic waves are totally reflected. This simplified model is a reasonable approximation for the propagation of sferics over distances smaller than a few hundered kilometers.

The ground with its finite conductivity, as well as the ionospheric layers, modifies the wave forms significantly at greater distances. We describe the properties of the wave guide between earth and ionosphere (the terrestrial wave guide) by a transmission function $T(\varrho, \omega)$ that depends on distance ϱ, frequency ω and other parameters (local time, season, latitude, geomagnetic field, etc). The vertical electric field of a pulse traveling in the wave guide is then modified (Wait 1970; Volland 1968) according to

$$E_z'(\varrho, t) = 1/(2\pi) \int_{-\infty}^\infty \hat{E}_z(\varrho, \omega) T(\varrho, \omega) \exp(-i\omega t) d\omega \; , \quad (7.9)$$

where

$$\hat{E}_z(\varrho, \omega) = \int_{\varrho/c}^\infty E_z(\varrho, t) \exp(i\omega t) dt \quad (7.10)$$

is the Fourier transform of the field over a perfectly conducting earth.

Considering only the radiation field of a vertical electric dipole [the last term on the right in Eq. (7.6)], Eq. (7.10) may be written as

$$\hat{E}_z(\varrho, \omega) = iM\alpha\beta\mu\omega / [4\pi\varrho(i\omega-\alpha)(i\omega-\beta)] \exp(ik\varrho) \; , \quad (7.11)$$

with $k = \omega/c$ the wave number in vacuum.

The maximum spectral amplitude occurs at the frequency

$$f_{max} = (\alpha\beta)^{1/2}/(2\pi) \; . \quad (7.12)$$

The transmission function of the terrestrial wave guide can be approximated by applying the two concepts of ray and mode theory. Ray theory describes the propagation of the pulse as the sum of a directly arriving ground wave and

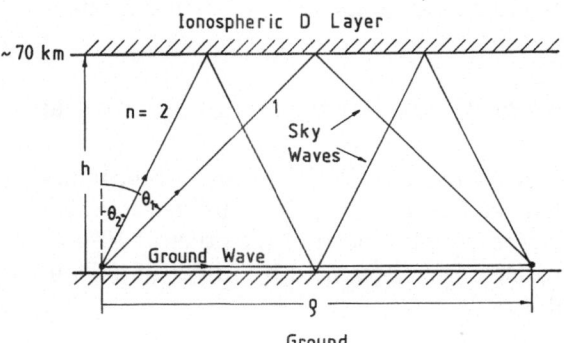

Fig. 7.7. Geometry of ray propagation within the terrestrial wave guide between earth and ionosphere. Ground wave and two sky waves are displayed

sky waves reflected at the base of the ionosphere near 70 to 90 km altitude depending on time of day and season:

$$T = T_0 + \sum T_n, \tag{7.13}$$

where T_0 is the transmission function of the ground wave, and

$$T_n \simeq R_i^n R_e^{n-1} [(1 + R_e)^2 / 2] \sin^3 \theta_n \exp[ik(r_n - \varrho)] \tag{7.14}$$

is the transmission function of the n-th hop sky wave (e.g., Volland 1968). R_i and R_e are the reflection factors of the ionospheric D-layer and the ground, respectively. $\theta_n = \arctan[\varrho/(2nh)]$ is the angle of incidence, $r_n = [\varrho^2 + 4n^2h^2]^{1/2}$ is the phase path and h is the virtual reflection height of the n-th sky wave (Fig. 7.7).

In the VLF range (very low frequencies: $3\,\text{kHz} < f < 30\,\text{kHz}$) and at distances $\varrho < 500$ km, one may use the approximations

$$T_0 \simeq 1; \quad R_e \simeq 1; \quad R_i \simeq -\exp(-2\alpha \cos\theta), \tag{7.15}$$

with

$$\alpha \simeq 0.001\,\omega^{2/3} \quad (\omega \text{ in s}^{-1}),$$

so that the sum of the ground wave and the first hop wave becomes

$$E_z' \simeq E_z(\varrho, t_r) - 2\sin^3\theta_1 \exp(-2\alpha\cos\theta)E_z(\varrho, t_r'), \tag{7.16}$$

with

$$t_r' = t_r - (r_1 - \varrho)/c.$$

The first hop wave arrives at a time $\Delta t = (r_1 - \varrho)/c$ later than the ground wave and has nearly the same wave form.

Pulses with pulse lengths $\tau < \Delta t$ can indeed be separated into a ground wave pulse and a first hop pulse at distances between about 200 and 500 km. This allows the determination of the source distance from the value of Δt (Taylor 1969; McDonnald et al. 1979).

The multihop components become more and more important at greater distances. In this case, it is more convenient to use the mode approximation of low frequency wave propagation. The transmission function of the n-th mode is given by (Wait 1970; Harth 1982)

$$T_n \simeq (1/h)[(\Theta/\sin\Theta)(2i\pi\varrho c/\omega)]^{1/2} K_n \exp[(-A_n+iB_n)\varrho] , \qquad (7.17)$$

with $\Theta = \varrho/a$ the polar distance between receiver and transmitter, a the earth's radius, K_n an amplitude factor, A_n an attenuation factor and B_n a phase factor.

In the VLF range, the ground behaves to a first approximation like an electric wall ($R_e \simeq 1$), and the ionosphere behaves like a lossy magnetic wall [R_i from Eq. (7.15)]. The eigenvalue equation of an plane wave guide is

$$R_i R_e \exp(2ikh\cos\theta_n - 2in\pi) = 1 , \qquad (7.18)$$

from which follows for the VLF range

$$\cos\theta_n \simeq \bar{C}_n[1+i\alpha/(kh)] , \qquad (7.19)$$

where

$$\bar{C}_n \simeq (n-1/2)\pi/(kh) \qquad (7.20)$$

are the eigenvalues of an ideal wave guide with $R_e = -R_i = 1$. In this ideal wave guide, the horizontal electric field disappears at the ground and the horizontal magnetic field disappears at the ionosphere (see Fig. 7.8).

In the ELF range (extremely low frequencies: $0.3\,\text{kHz} < f < 3\,\text{kHz}$), the ionosphere behaves like an electric wall ($R_i \simeq 1$). Therefore,

$$\cos\theta_n \simeq \bar{C}_n \simeq n\pi/(kh) . \qquad (7.21)$$

The phase factor B_n in Eq. (7.17) is related to $\cos\theta_n$ by

$$B_n = k[(1-\cos^2\theta_n)^{1/2}-1] \simeq -k\bar{C}_n^2/2 . \qquad (7.22)$$

It can be shown that the attenuation factors A_n increase with wave number n, and that the zeroth mode dominates in the ELF range and the first mode dominates in the VLF range at large distances.

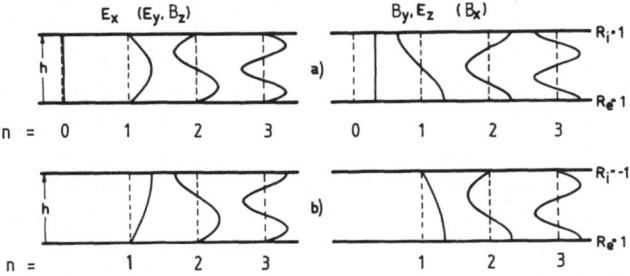

Fig. 7.8a, b. Vertical mode structure of electric (E) and magnetic (B) field components within an ideal plane wave guide with an electric wall at the bottom ($R_e = 1$) and a magnetic wall at the top ($R_i = -1$) simulating the terrestrial wave guide for VLF propagation (*lower panel*), and wave guide with two electric walls simulating ELF propagation (*upper panel*). The first four modes are displayed

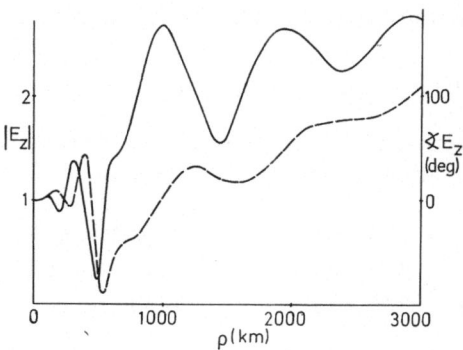

Fig. 7.9. Normalized vertical electric field versus distance in amplitude (*solid line*), and phase (*dashed line*). The transmitter is a vertical dipole radiating at 15 kHz. The virtual reflection height of the terrestrial wave guide is 70 km corresponding to daytime conditions at midlatitudes. The interference minimum near 500 km distance is the last interference minimum of ground wave and first sky wave. It is also the first interference minimum between first and second mode

The equivalence between ray theory and mode theory can be seen immediately from a comparison between the interference pattern in the VLF range of ground wave and first hop wave on the one hand, and first and second mode on the other hand (Fig. 7.9). The location of interference minima of the two first rays is at [see Eqs. (7.13), (7.14) and (7.15)]

$$k(r_1 - \varrho) - \pi = (2m - 1)\pi \quad (m = 1, 2, \ldots) \tag{7.23}$$

or at a distance $\varrho_m \simeq k h^2 / (m\pi)$. The interference minima of the two first modes is at [see Eqs. (7.17), (7.20) and (7.22)]

$$(\bar{C}_2^2 - \bar{C}_1^2) k\varrho / 2 = (2m' - 1)\pi \quad (m' = 1, 2, \ldots) \tag{7.24}$$

or at $\varrho_{m'} = (2m' - 1) k h^2 / \pi$. The last interference minimum of ray theory is for $m = 1$, or at $\varrho_1 = k h^2 / \pi$. This is also the location of the first interference minimum of mode theory ($m' = 1$). The equally spaced interference minima at $\varrho \gtrsim 500$ km of the vertical electric field of a 15 kHz transmitter in Fig. 7.9 are due to the interference of the two first modes. The interference minima at $\varrho \lesssim 500$ km are due to the ground wave and the one hop wave. Ray theory is, therefore, a good approximation at distances $\varrho < \varrho_1$ where only the ground wave and the first hop sky wave are important. Mode theory is appropriate at larger distances where only the two first modes are significant.

Figure 7.10 shows the magnitudes $|T_0|$ and $|T_1|$ of the transmission functions of the zeroth and the first mode as a function of frequency and distance. The zeroth mode is least attenuated at frequencies <2 kHz. The first mode has its attenuation minimum between about 5 and 20 kHz, depending on distance. The transition range between zeroth and first mode is near 2 kHz where the attenuation attains a relative maximum. The wave guide thus has two windows, one in the ELF range, the other one in the VLF range.

The spectral field strength $|\hat{E}_z T|$ from Eqs. (7.11), (7.13) and (7.17) is plotted in Fig. 7.11 versus frequency with the distance as parameter for the type 2 return stroke R_3 from Table 6.2. Since $T \simeq 1$ for $f < 100$ kHz at a distance of $\varrho < 200$ km, the upper curve in Fig. 7.11 is the magnitude of $\hat{E}_z$ in Eq. (7.11). The difference between the solid and the dashed curve indicates the attenuation

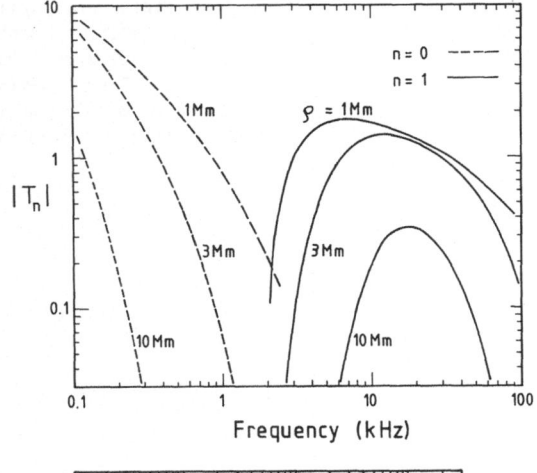

Fig. 7.10. Magnitude $|T_n|$ of transmission functions of the zeroth and the first mode versus frequency at distances 1000, 3000, and 10,000 km during daytime conditions at midlatitudes

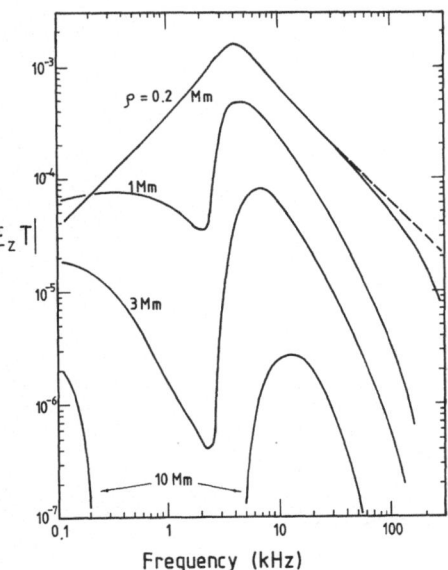

Fig. 7.11. Spectral field strength (magnitude) of type 2 return stroke R_3 from Table 6.2 versus frequency at different distances. *Dashed curve* gives the spectral amplitude at 200 km distance in the case of a perfectly conducting ground

of the ground wave at frequencies >100 kHz. The band pass effect shifts the maximum of the spectral amplitude from 4 kHz at distances <200 km to 12 kHz at 10,000 km. More detailed calculations of the transmission function as a function of local time, geomagnetic field, curved earth, etc. have been made by Harth (1982).

Figure 7.12 shows sferic wave forms observed at successively greater distances from the source. One notices the filter effect by the suppression of frequencies outside the VLF range.

The phase velocity of the modes, which is

$$v_p = c/(1 + B_n/k) \tag{7.25}$$

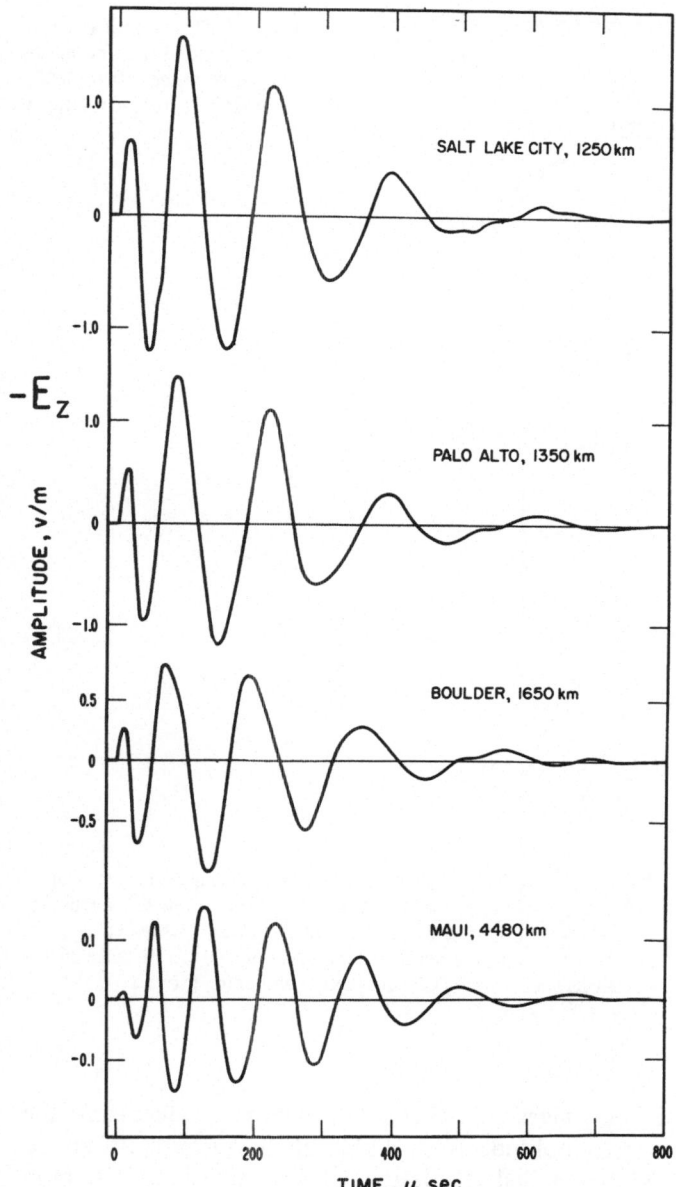

Fig. 7.12. Typical wave forms of vertical electric field from sferics at great distances from the source (Taylor 1960). Orientation of electric field is positive downward (Fig. 14.1b)

decreases with increasing frequency and is larger than the speed of light within the VLF range [Eq. (7.22)]. This frequency dependence can be used to locate the sources of sferics from greater distances (>1000 km) by measuring the difference in arrival time of the pulse trains at two or more frequencies (Sao and Jindoh 1974; Volland et al. 1983).

The transmission function of the terrestrial wave guide has a window at extremely low frequencies. Pulses containing significant spectral amplitudes in that frequency range may be received around the earth. The spectral amplitudes of lightning pulses, being proportional to f at frequencies $f \ll f_{max}$ [see Eqs. (7.11) and (7.12)], are generally small in the ELF range. Occasionally, pulses with stronger spectral amplitudes in the ELF range are observed as wave forms with a low frequency tail [slow tail sferics (Taylor and Sao 1970)].

7.4 Schumann Resonances

Lightning signals in the frequency band $f < 100$ Hz are very weak, and, moreover, are disturbed by man-made noise (electric power lines, etc.). However, the terrestrial wave guide behaves like a resonator for the zeroth order mode in that frequency band. Since the idealized zeroth order mode has only a vertical electric component which is constant with height (see Fig. 7.8) and a phase velocity which equals the speed of light, resonance occurs for waves with horizontal wavelengths which are an integral multiple of the earth's circumference. This yields a resonance frequency of

$$f_m = mc/(2\pi a) = 7.5\, m\, \text{Hz} \quad (m = 1, 2, 3, \ldots) . \tag{7.26}$$

The first resonance peaks are at 7.5, 15, 22.5, and 30 Hz. The spectral signals from lightning are amplified at these frequencies.

A more sophisticated treatment (Polk 1982) accounts for the spherical earth and the influence of the lossy and anisotropic ionospheric boundary. It turns out that the resonance condition (7.26) is changed by replacing m with $[m(m+1)]^{1/2}$ in the case of an ideal wave guide.

The first resonance is predicted to occur now at $f_1 = 10.6$ Hz. However, the lossy wave guide reduces the observed resonance frequencies to the values indicated in Fig. 7.13.

The horizontal configuration of the resonance modes is fixed by the location of the source and is shown in Fig. 7.14 for the vertical electric field E_r and the azimuthal magnetic field B_λ (in spherical coordinates) of the two first

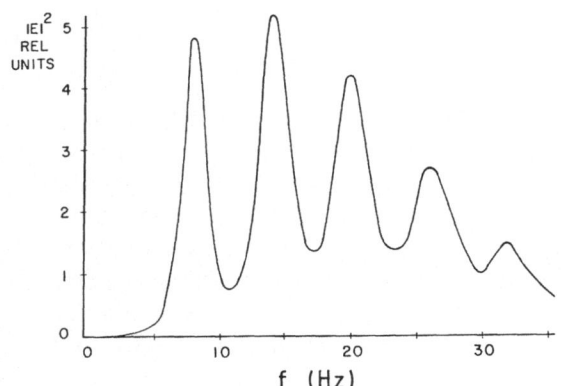

Fig. 7.13. Amplitude spectrum of electric field obtained at Kingston, R. I., showing the first five Schumann resonances. (Polk 1982)

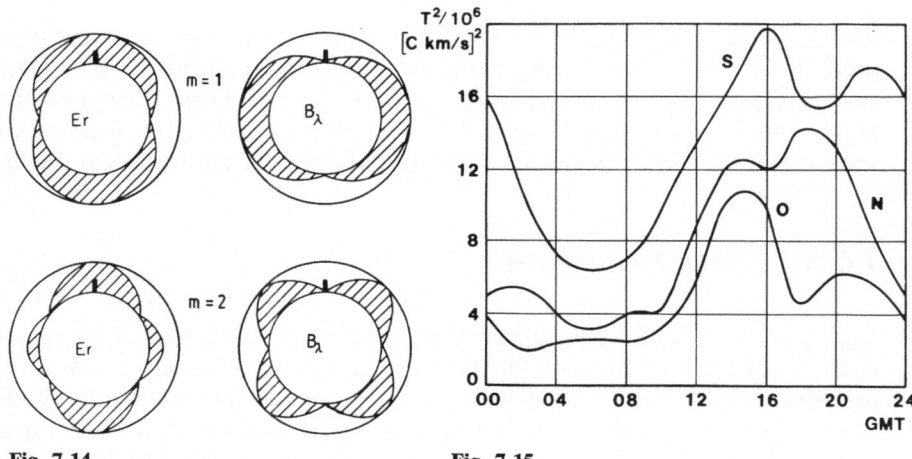

Fig. 7.14   **Fig. 7.15**

Fig. 7.14. Idealized amplitude distribution of radial electric field and azimuthal magnetic field of first and second Schumann resonance over a perfectly conducting ground, assuming a single source at the *thick vertical dash*. (Polk 1982)

Fig. 7.15. Diurnal variation of lightning activity, derived from Schumann resonance data in September (S), October (O), and November (N), 1970 (Polk 1982). The parameter T (in $A^2 km^2$) is a measure of total charge transfer during the lightning events

resonance modes. The real ionosphere, which varies not only in height but also in latitude and longitude, modifies the mode structure.

The activity of Schumann resonances reflects the global lightning activity. Since three main thunderstorm areas exist (Africa, South and Middle America, East Asia) with their daily variations peaking at local afternoon, one can use the latitude dependence of the resonance modes from Fig. 7.14 to evaluate global thunderstorm activity from one station. Figure 7.15 shows the diurnal variation of global lightning activity derived from Schumann resonances during fall 1970. Maximum activity occurs between 1600 h and 2000 h GMT, which is consistent with the results in Figs. 3.3c and 4.1.

7.5 Nuclear Electromagnetic Pulses

Explosions of atomic bombs generate electromagnetic pulses (EMP). Since these pulses have certain similarities with sferics (e.g., Uman et al. 1982), we discuss in this section the phenomenon of nuclear EMP's. During a nuclear explosion, an energy of 4.2×10^{15} J per megaton is released within a few nanoseconds. The energy of a one megaton nuclear device is the equivalent of one million tons of TNT. About 0.1% of this energy goes into prompt gamma rays and about 1% into fast neutrons. Secondary gamma rays are produced by fast neutrons with energies >5 MeV through inelastic collisions with the air molecules. For bursts near the ground, inelastic collisions of the fast neutrons with the ground are a further source of secondary gamma rays.

Gamma rays in the MeV-range easily expel electrons from their atomic and molecular bonds by the Compton effect. Each 1 MeV gamma ray eventually produces about 30,000 electron-ion pairs via a cascade process. This ionization process is completed after a few nanoseconds at sea level and within about 50 ns at 30 km altitude. The forward scattered Compton recoil electrons move much faster than the ions. Therefore, a radial electric current, the so-called radial Compton current, flows toward the source and is accompanied by a radial electric field. On the other hand, the thermalized component of the recoil electrons drastically increases the electric conductivity of the air.

The temporal variation of the radial electric field is similar to that of the electric field during the noninductive electrification in thunderclouds [see Eq. (5.4) and (Longmire 1978)]:

$$\varepsilon \, \partial E_r / \partial t + \sigma E_r + j_r = 0 \, , \tag{7.27}$$

where j_r is the radial Compton current, σ is the enhanced electric conductivity, and E_r the radial electric field. The electric field saturates at a value close to

$$\hat{E}_{max} \simeq -j_r / \sigma \lesssim 50 \, \text{kV/m} \tag{7.28}$$

because σ and j_r are both proportional to the dose rate of the gamma rays and to the density of the air. Thus, the saturation field does not greatly depend on the strength of the explosion. This saturation electric field remains for about 10 ns for bursts at sea level and somewhat longer for explosions at greater heights.

As long as the Compton current is spherically symmetric, no magnetic field and, therefore, no electromagnetic pulse is generated. If the burst is on or near the ground, however, the original radial field is short-circuited on the ground, and an electromagnetic field pattern is created similar to that of a vertical electric dipole with a field configuration as in Eqs. (7.6) and (7.7).

A typical wave form of the radial Compton current is shown in Fig. 7.16. It is proportional to the prompt gamma ray flux and can be approximately described by the Bruce-Golde formula (6.1) with the parameters

$$\alpha_r \simeq 3.5 \times 10^7 \, \text{s}^{-1}; \quad \beta_r > 3.5 \times 10^9 \, \text{s}^{-1}; \quad f_{max} > 56 \, \text{MHz} \, ,$$

where f_{max} from Eq. (7.12) is the maximum spectral frequency of the electromagnetic pulse.

The electromagnetic energy of ground bursts reaches saturation electric fields within a distance of about 10 km from the source. Outside this range, the energy propagates like that of a vertical electric dipole. This energy is attenuated in the terrestrial wave guide as shown in Fig. 7.10, so that the high frequency component is strongly suppressed after a distance of a few 100 km from the source. Only the low frequency component ($f < 100$ kHz) remains with a wave form similar to sferic wave forms (Price 1974).

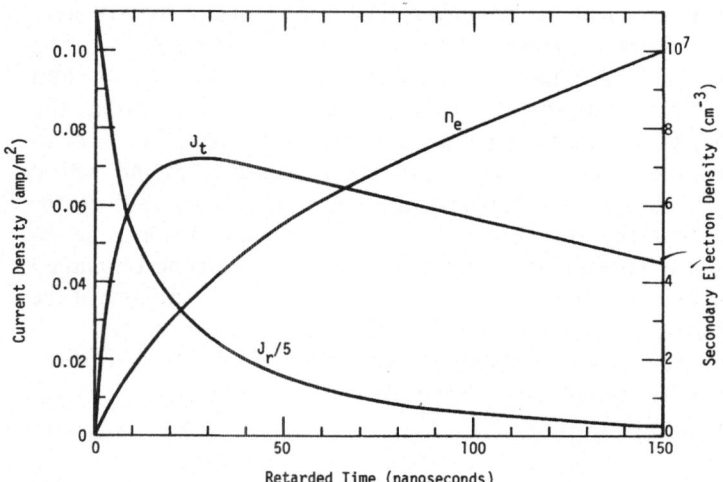

Fig. 7.16. Typical wave form of radial Compton current J_r during a nuclear explosion on the ground, and transverse Compton current J_t during a high altitude nuclear explosion. Also shown is the increase of electron number density n_e with time in the environment of a ground-based explosion. (Longmire 1978)

The geomagnetic field forces the recoil electrons to spiral along the geomagnetic lines of force. The Larmor radius of 1 MeV electrons is of the order of 100 m at midlatitudes. At low altitudes, the mean forward range of the recoil electrons before scattering is a few meters, so that the electrons never complete a full Larmor circle. At 30 km altitude, the mean forward range becomes comparable to the Larmor radius, so that the gyrating electrons also generate a transverse Compton current that radiates electromagnetic energy like a current loop.

The prompt gamma rays from bursts at high altitudes, propagating with the speed of light, produce Compton recoil electrons at heights from 50 to 30 km. These electrons gyrate in the geomagnetic field and generate synchronous transverse Compton currents in an area determined by the optical horizon of the burst. A typical wave form of a transverse Compton current is shown in Fig. 7.16. It can be simulated by the Bruce-Golde formula (6.1) with the parameters

$$\alpha_t \simeq 4.5 \times 10^6 \, \text{s}^{-1}; \quad \beta_t \simeq 1.1 \times 10^8 \, \text{s}^{-1}; \quad f_{\text{max}} \simeq 3.5 \, \text{MHz} \,.$$

The transverse Compton currents act like a huge aerial antenna system from which a plane electromagnetic wave covering an area of serveral 1000 km in horizontal extent propagates in the direction of the geomagnetic field lines down to the ground. This wave has a pulse form with maximum spectral frequency in the 1 – 10 MHz band and maximum amplitudes of the order of 50 kV/m [Eq. (7.28)].

It is the electromagnetic pulse of this transverse Compton current, which excites resonant oscillations on compact technical systems with metallic envelopes (cars, airplanes, etc.). Such systems react like receiving antennas. For example, the radial electric field of the fundamental eigenfrequency of a highly conducting sphere with radius R (Sommerfeld 1952) is

$$E_r = E_0 \sin(\omega_r t) \exp[-ct/(2R)] , \tag{7.29}$$

where $f_r = \omega_r/(2\pi) = 1.73\,c/(4\pi R)$ is the resonance frequency of the sphere. Equation (7.29) should be compared with the field of a thin wire as an example of the eigenmode of an elongated configuration with finite conductivity [Eq. (6.16)].

A sphere of diameter $R = 10$ m has a fundamental resonance frequency at $f_r = 4.1$ MHz, which lies well within the spectral range of nuclear electromagnetic pulses. Thus, metallic covers of technical systems reradiate high frequency pulses with large peak amplitudes. These pulses may destroy sensitive electronic equipment by electromagnetic induction effects (Baum 1978).

II Tidal Wind Interaction with Ionospheric Plasma

8 Electric *Sq* and *L* Currents

We shall discuss in the next two chapters the second generator of electromagnetic energy in the atmosphere: the ionospheric dynamo. The dynamo coil is the electrically conducting air within the dynamo region between about 80 and 200 km height. The driving force is the tidal wind which moves the ionospheric plasma against the geomagnetic field and induces electric fields and currents. Manifestations of the electric currents on the ground are regular variations of the geomagnetic field depending on solar day (*Sq* = solar quiet) and lunar day (*L* = lunar), respectively. In this chaper, we review the observations. In the next chapter, we deal with the so-called dynamo theory of the *Sq* current.

8.1 Origin of *Sq* and *L*

The solar tidal winds are excited by solar differential heating of the atmosphere which is accompanied by day–night differences in atmospheric pressure and temperature. The winds are set into motion in order to compensate for these pressure differences. The basic period of the solar tides is one solar day. The basic wavelength is the size of the earth. Tides can, therefore, only exist in form of individual wave modes the horizontal structure of which depends on the spherical and rotating earth. For more details, see Sect. 9.1. The heat source is solar radiation which is absorbed by water vapor insolation in the lower atmosphere, by exciting neutrals in the middle atmosphere, and by ionizing the neutrals within the upper atmosphere.

Two atmospheric wave types exist within the lower and middle atmosphere: internal or travelling waves which can transport wave energy, and external or evanescent waves which cannot transport energy. The second wave type has infinitely large vertical wavelengths, meaning that its phase remains constant with height, and that the wave amplitude decreases exponentially with height outside the source region. The fundamental solar tidal wave is the diurnal $(1, -2)$ mode (for the nomenclature, see Sect. 9.1). It belongs to the external waves. On the other hand, the solar semidiurnal $(2, 2)$ mode is an internal wave, apart from a height region near the mesopause (Chapman and Lindzen 1970). The difference between internal and external waves disappears above about 150 km altitude because of the increasing influence of wave dissipation.

These regular tidal winds drive the ionospheric plasma at dynamo layer heights. However, ions and electrons are affected differently by the winds (see Sect. 2.4), so that a horizontal electric current flows. Electric charge separation sets up an electric polarization field. The magnetic field generated by this current can be measured on the ground as geomagnetic Sq variation, superposed on the geomagnetic field from the earth's interior. A narrow band of strong currents flows near the magnetic equator – the equatorial electrojet – resulting from the enhanced electric conductivity (the Cowling conductivity; see Fig. 2.5) in that region.

It is the solar (1, −2) diurnal tide with its nearly constant phase which is most effective in generating synchronous electric currents within the whole dynamo region (the Sq current) so that their magnetic effect is observable. The internal tidal modes, in particular the solar semidiurnal (2, 2) mode, are not so effective because of their finite vertical wavelengths. They contribute mainly to the day-to-day variability of Sq on the ground.

Atmospheric lunar tides are excited by the gravitational force which the moon exerts on the earth. The fundamental lunar tide is the semidiurnal (2, 2) mode. Its pressure amplitude on the ground is smaller by a factor of about 20 as compared with the pressure amplitude of the solar semidiurnal tide. The electric current induced by the lunar tide at dynamo region heights, called the L current, is smaller than the Sq current by the same order of magnitude.

8.2 Observations of Sq Variations on the Ground

About 99% of the geomagnetic field on the ground is generated within the earth's core. Its dominant component appears as an inclined magnetic dipole field at the earth's surface (see Sect. 14.2). Superposed on this nearly constant core field are shorter periodic variations ranging from a fraction of a second to years. This component comprising not more than 1% of the total field has its origin in electric current systems flowing outside the earth in the ionosphere and magnetosphere at heights above about 100 km.

Figure 8.1 shows two typical magnetograms of geomagnetic variations at midlatitudes versus local time. The "quiet" day (Fig. 8.1a) reflects mainly the regular variations (Sq) while the "disturbed" day (Fig. 8.1b) displays a superposition of magnetic fields from irregular magnetospheric currents on Sq. In order to separate the Sq variations from these disturbed fields, one has to select "quiet" days and eliminate the residual disturbances by the method of superposed epochs.

Average Sq variations versus local time are shown in Fig. 8.2 for various latitudes. It is convenient to ascribe such magnetic variations to an equivalent electric overhead current assumed to flow in a thin spherical shell near 100 km height. The horizontal magnetic component ΔH_p of a homogeneous infinitely extended electric current of current density j flowing in a thin sheet of thick-

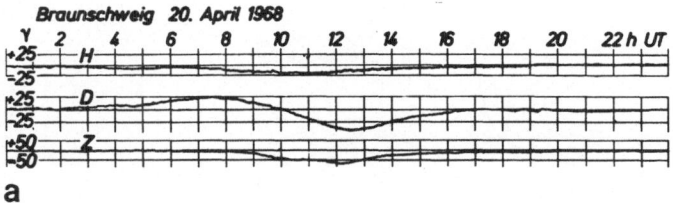

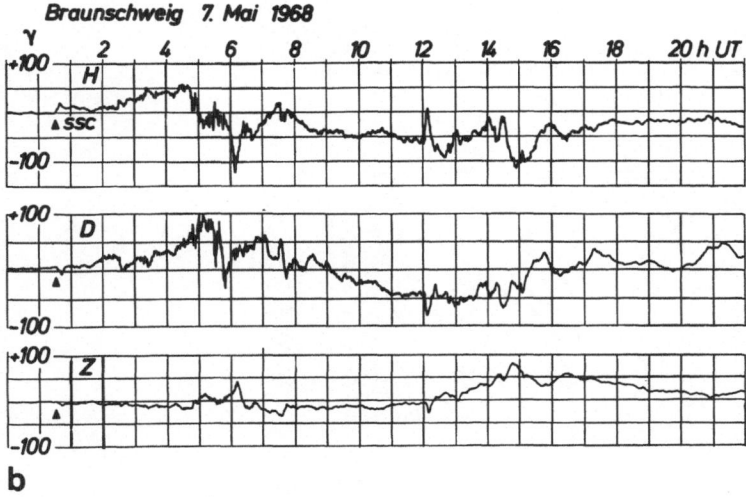

Fig. 8.1. Typical magnetograms of geomagnetic variations of horizontal intensity ΔH, declination ΔD, and vertical intensity ΔZ during quiet (**a**), and disturbed (**b**) conditions (Kertz 1969). (Note: $1\,\gamma = 1\,\text{nT}$)

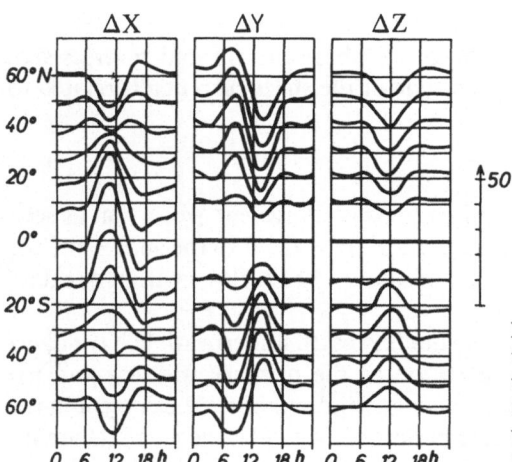

Fig. 8.2. Average geomagnetic *Sq* variation of north (ΔX), east (ΔY), and vertical (ΔZ; positive downward; see Fig. 14.2) component versus local time at various latitudes. (Chapman and Bartels 1951)

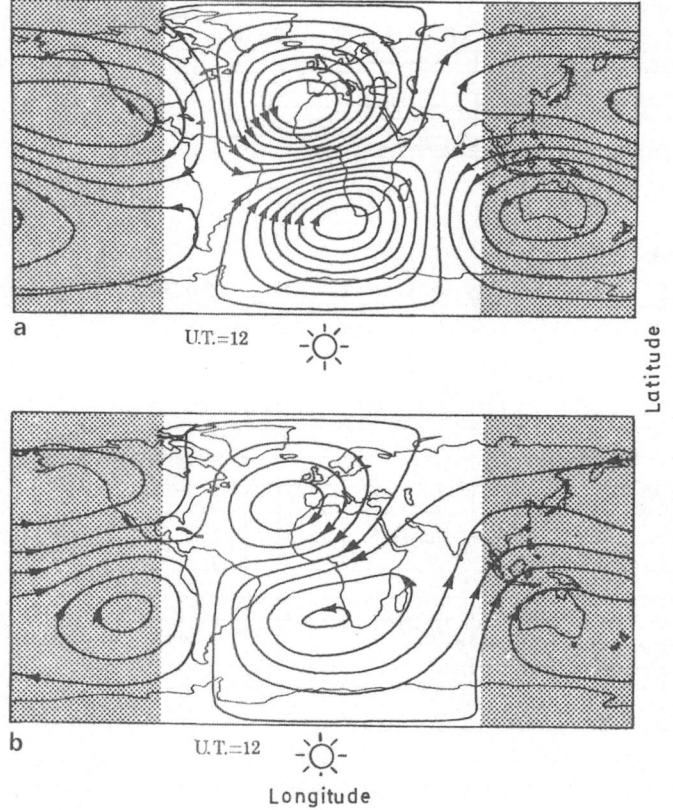

Longitude

Fig. 8.3a, b. Stream lines of equivalent ionospheric Sq current during equinox (1957–1960) at 1200 UT separated into primary (**a**), and secondary (**b**) part. Between two stream lines flow 20 kA. (Malin 1973)

ness Δh is orthogonal to the direction of the sheet current and reverses sign above the sheet (Fig. 14.3c). The magnitude of the sheet current J is related to ΔH_p as [Eq. (14.25)]

$$J = j\Delta h \simeq 2\Delta H_p/\mu .\tag{8.1}$$

Figure 8.3a shows current stream lines of this equivalent Sq current as seen from the sun at noon. This current configuration is fixed to the sun, while the earth rotates beneath it. A total current of about 140 kA flows within one daytime vortex.

The rotating Sq current and the conducting earth underneath behave like a huge transformer with the dynamo region as the primary winding and the electrically conducting earth as the secondary winding. Since the current in the primary winding varies with the basic period of 24 h, electric currents are induced in the earth's interior (e.g., Price 1967). The magnetic field of this secondary current is superposed on the magnetic field of the primary dynamo current. Methods to separate both components go back to Gauss (e.g., Malin 1973). Figure 8.3b gives the secondary current system within the earth. Its amplitude is about 1/3 of that of the primary current and is slightly shifted in phase. The ratio between the magnetic horizontal component on the ground and its corresponding equivalent overhead current is, therefore, instead of Eq. (8.1)

$$\Delta H/J \simeq 2\mu/3 \,. \tag{8.2}$$

The *Sq* current depends on season. The summer vortex is intensified compared with the winter vortex and reaches into the winter hemisphere. A longitudinal dependence of the *Sq* current exists which is related to the inclined dipole component of the internal geomagnetic field (Matsushita 1967).

One can develop the observed *Sq* variations into a Fourier series which yields, e.g., for the horizontal component of the geomagnetic variation,

$$\Delta H = \sum_m h_m(\phi) \cos[m\tau + \gamma_m(\phi)] \tag{8.3}$$

where ϕ is the geomagnetic dip latitude, and τ the local time. h_m and γ_m are amplitude and phase, respectively, of the m-th Fourier component. The geomagnetic dip latitude is defined by

$$2\tan\phi = \tan I \tag{8.4}$$

where I is the dip angle of the geomagnetic field vector at each station. For an ideal inclined dipole, ϕ is identical with the geomagnetic latitude [Eq. (14.7)].

Figure 8.4 shows amplitude h_m and phase γ_m as functions of dip latitude for the first four Fourier components of ΔH during equinox derived from stations in the North–South American zone. Remarkable are the peak amplitudes at the geomagnetic dip equator which are produced by a narrow current band – the equatorial electrojet (e.g., Forbes 1981). The first Fourier component dominates. The shapes of the three other components are very similar to that of the first component. The amplitudes of the first three components behave roughly like $4:2:1$. A phase shift of about 180° occurs near the focii of the vortices. The first and second component are out of phase by about 180°. The zeroth Fourier component ($m = 0$) cannot be determined uniquely because the baseline of the geomagnetic variations is not known. In general, one assumes as baseline the average of the night time amplitudes.

The centers of the daytime vortices show a day-to-day variability in latitude as well as in longitude. This can be attributed in part to the influence of the internal tidal wave modes which are sensitive to the varying meteorological conditions in the lower and middle atmosphere, and in part to solar activity (Sect. 8.5) (Brown 1975).

The electric field of *Sq* mapped down into the lower atmosphere has not yet been identified uniquely. Its amplitude is expected to be of the order of 1 V/m at midlatitudes on the ground (Sect. 9.7), too small to be detectable within the highly fluctuating thunderstorm field (see Fig. 3.4).

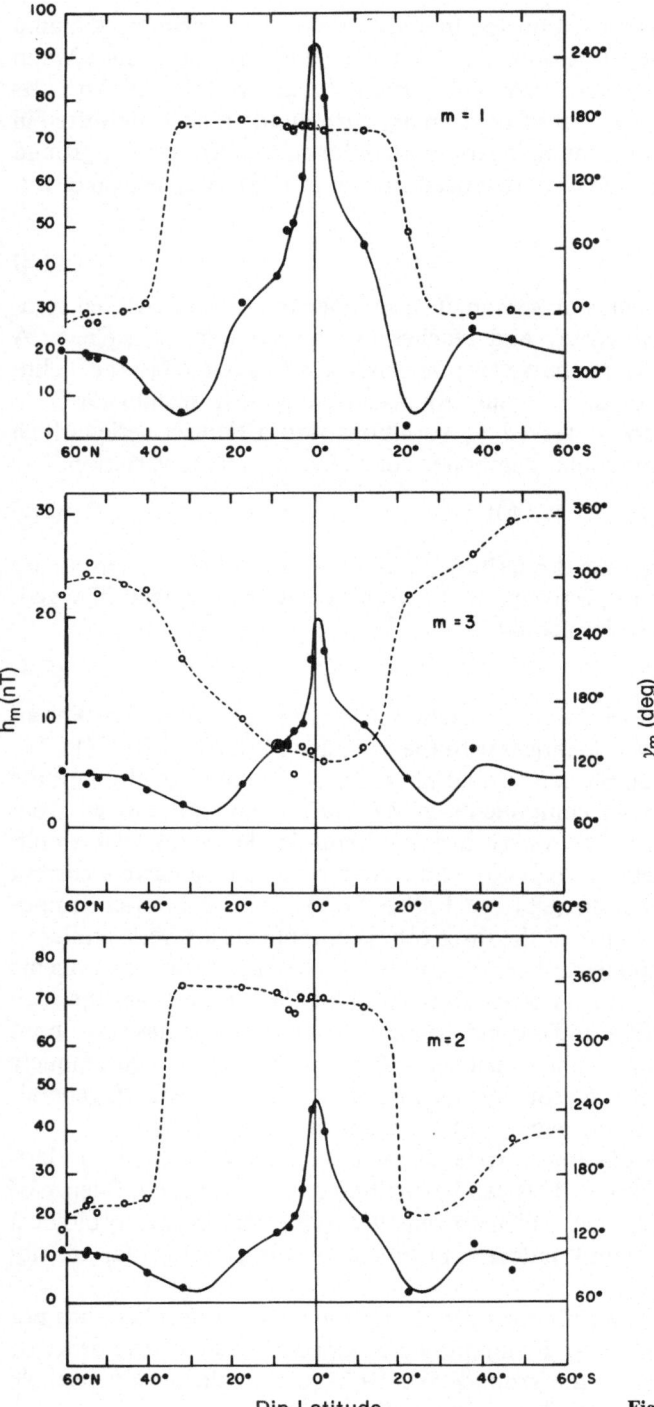

Dip Latitude

Fig. 8.4

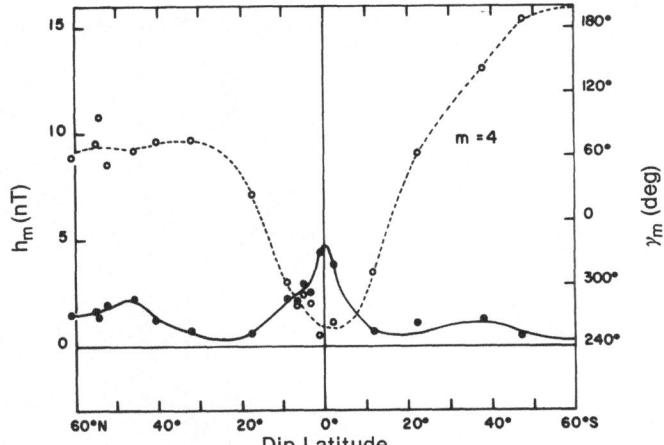

Fig. 8.4 (continued)

Fig. 8.4. Dip-latitude distribution of amplitudes h_m (*solid circles*), and phases γ_m (*open circles*) of the first four Fourier harmonics for the horizontal intensity ΔH of the geomagnetic *Sq* variation determined from data in the American sector during equinox 1958. *Solid curves* represent smoothed values. (Matsushita 1967)

8.3 Morphology of *L*

Lunar (*L*) variations are much weaker in amplitude than *Sq* variations. Isolation of such a weak signal is possible only by an appropriate averaging method of superposed epochs in terms of lunar time (Chapman and Bartels 1951). The dominant Fourier component is the lunar semidiurnal wave. Figure 8.5 shows the yearly average of the equivalent ionospheric *L* current, separated into its primary and secondary parts.

The seasonal variation of the *L* field is very similar to that of the *Sq* field. The intensity of the primary *L* current is about 1/20 of that of the *Sq* current. During sunlit hours, the *L* current is strongly enhanced, while it approaches zero during the night. Thus, the *L* field exhibits, in addition, a solar lunar modulation depending on the lunar phase. This can clearly be seen in Fig. 8.6 which gives the declination ΔD of *L* measured in Djarkarta, Indonesia, at various lunar phases, and averaged over all lunar phases.

8.4 Space Observations Related to *Sq*

The *Sq* current manifests itself by its magnetic field in space above and below the dynamo layer. The *Sq* current is also accompanied by an electric field which is maintained by charge separation between ions and electrons.

Measurements of electric and magnetic fields in space related to *Sq* have been made on board satellites and rockets. Figure 8.7a shows the result of a

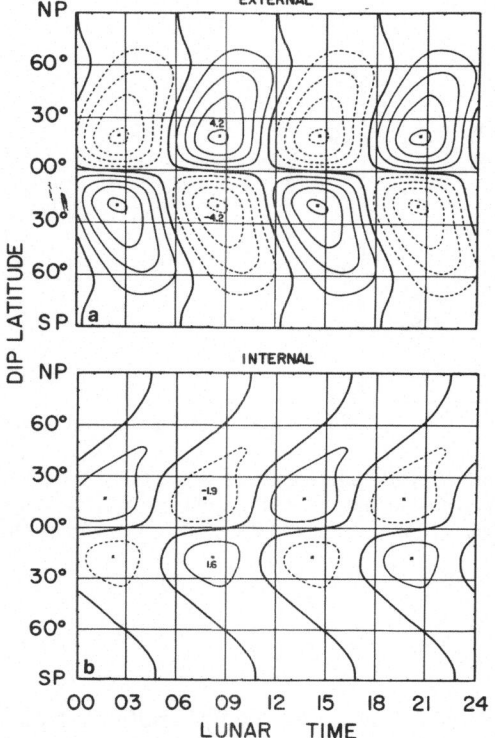

Fig. 8.5 a, b. Primary (a), and secondary (b) part of equivalent ionospheric *L* current system (yearly average) versus lunar local time and dip latitude. Between two lines flow 1 kA. *Thick solid curves* indicate zero-intensity lines. *Thin solid* and *dashed lines* show counterclockwise and clockwise currents, respectively. (Matsushita 1967)

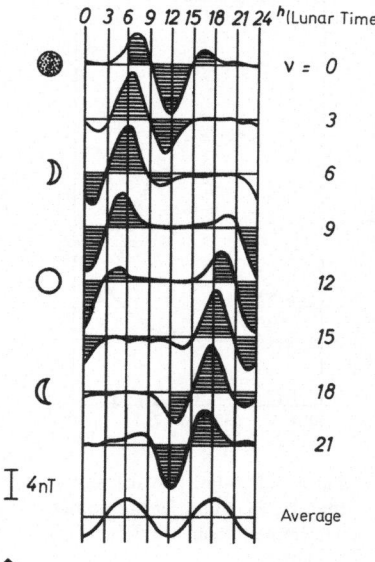

Fig. 8.6. *L* variation of geomagnetic declination ΔD in Djarkarta, Indonesia, (6°S) at eight different lunar phases, and average over all lunar phases. *Hatched areas* indicate daylight hours. (Kertz 1969)

rocket flight: the residual north component ΔX of the geomagnetic field is plotted as function of height near the geomagnetic dip equator. The equatorial electrojet, which is a small band of strong currents flowing to the east near 110 km altitude, behaves to a first approximation like a current sheet of thickness $\Delta h \simeq 20$ km. The north component $\Delta X \simeq \pm 100$ nT above and below the sheet in Fig. 8.7a yields a current strength of $J \simeq 0.16$ A/m [Eq. (8.1) and Fig. 14.3c] or an average current density of $j = J/\Delta h \simeq 8\ \mu$A/m². Figure 8.7b shows the height profile of this current density as derived from Fig. 8.7a.

Rocket observations of the magnetic effect of the *Sq* current outside the equatorial region are not so convincing. One measures a small discontinuity near 115 km, probably related to the Hall current component (Burrows and Hall 1965; Davis et al. 1965). The Pedersen current, however, which flows in a broader region near 140 km (see Fig. 1.1) has not been detected magnetically.

Magnetic measurements on board the NASA spacecraft MAGSAT indicate that the satellite passes through a meridional electric current system

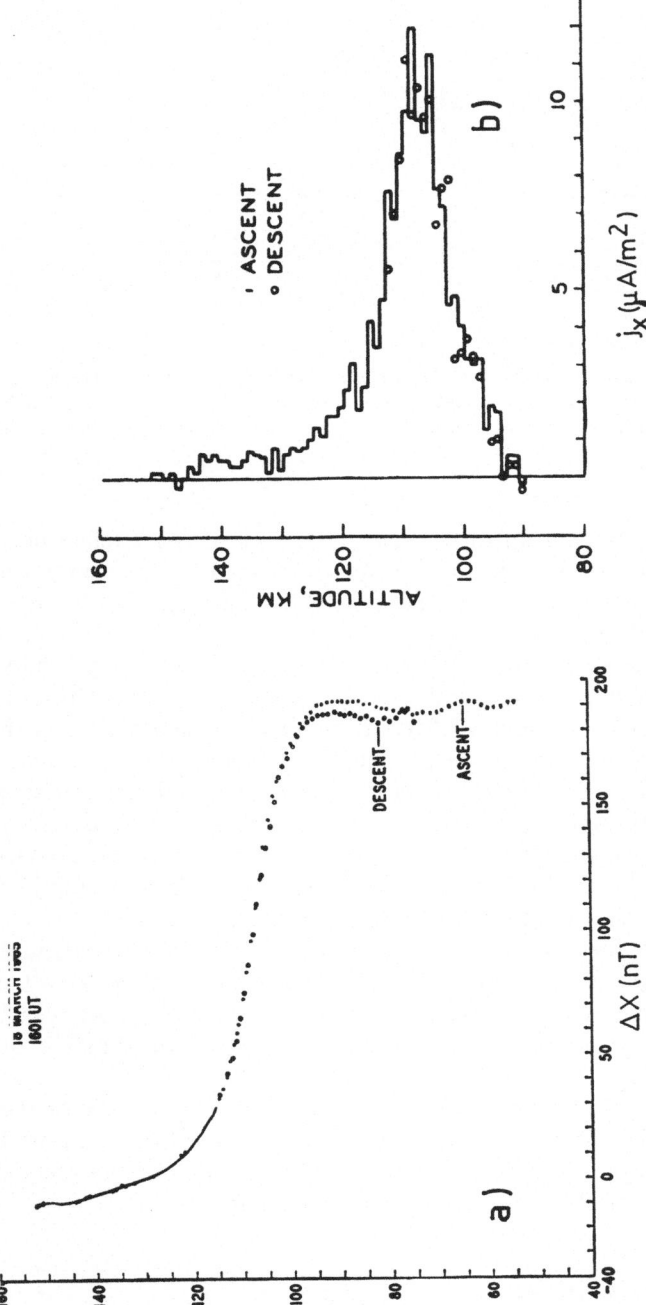

Fig. 8.7. a Rocket observations of north component ΔX of the equatorial electrojet at the geomagnetic equator during noon as function of height. Data of ascent and descent are separated. **b** Eastward electric current density of the electrojet versus height, as derived from the magnetic field measurements in Fig. 8.7a. (Davis et al. 1967)

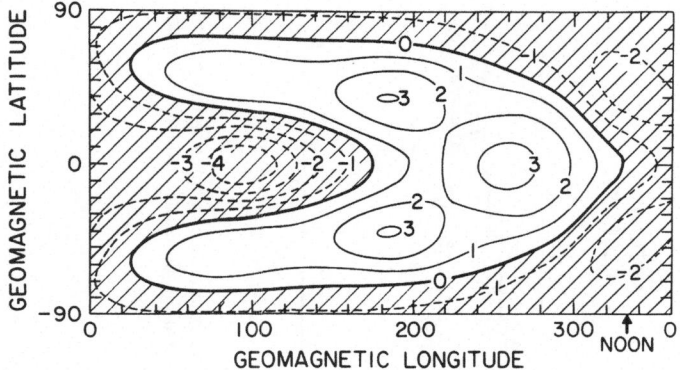

Fig. 8.8. Equipotential lines of the potential of the horizontal electric field at ionospheric heights plotted versus geomagnetic latitude and longitude (or geomagnetic local time) during quiet conditions, as derived from backscatter radars. Between two lines is a voltage difference of 1 kV. (Richmond 1976)

reaching from about 100 km up to at least the apogee of the satellite near 500 km and extending from 8° north to 8° south. No magnetic counterpart has been detected on the ground (Maeda et al. 1982). This meridional current system is related to the equatorial electrojet (see Sect. 9.6).

Rocket measurements of the electric field related to *Sq* give only "snapshots" and are not very useful for deriving the global field pattern. Ground-based backscatter radars transmit VHF-UHF pulses (30 – 3000 MHz) which are reflected at inhomogeneities of the plasma at ionospheric *F* layer heights. From the Doppler shift of the reflected signal, one can determine the plasma drift orthogonal to the geomagnetic field lines. Assuming low conductivity transverse to the geomagnetic field lines, the plasma drift velocity is related to the electric field as in Eq. (2.13) (Evans 1978). The electric field so determined is believed to be a quasi-static field, related to an electric potential.

Figure 8.8 shows lines of equal electric potential (electric equipotential lines) versus geomagnetic local time and latitude derived from backscatter radars during geomagnetically quiet days. The electric field pattern below 60° latitude is believed to be predominantly due to the electric polarization field of the *Sq* current.

From VHF-UHF radar measurements, one observes irregularities of the electron density in the regions of maximum current density of the equatorial electrojet. These irregularities result from instabilities of the ionospheric plasma (e.g., Crochet et al. 1979; Fejer and Kelley 1980).

8.5 Solar Influences on *Sq*

Sq variations are caused by the coupling between atmospheric tidal waves and the ionospheric plasma. The larger part of the tidal wave energy is generated by the rather constant visible and near-ultraviolet solar radiation which is absorbed within the lower and middle atmosphere. The day-to-day variability of the tidal waves is primarily a meteorological problem. Tidal waves propagate within the fluctuating background wind, which itself depends on the unstable conditions in the troposphere (e.g., Forbes 1982a, b).

On the other hand, the ionospheric plasma is primarily maintained by the variable X-ray and extreme ultraviolet (XUV) radiation from the sun, which is absorbed at thermospheric heights above 100 km. Any change of the solar input will then modulate the electric conductivity, and by that the *Sq* current. The diurnal variation of the electric conductivity resulting from the day−night contrast is mainly responsible for the existence of the higher harmonics in *Sq*. This can be seen immediately from a simple estimate. The predominant solar tide generating the *Sq* current is a diurnal wave depending on solar local time τ. Its horizontal velocity is given by $U \simeq U_1 \cos \tau$. The electric conductivity also varies with local time like $\sigma \simeq \sigma_0 - \sigma_1 \cos \tau$. The *Sq* current is proportional to the product σU:

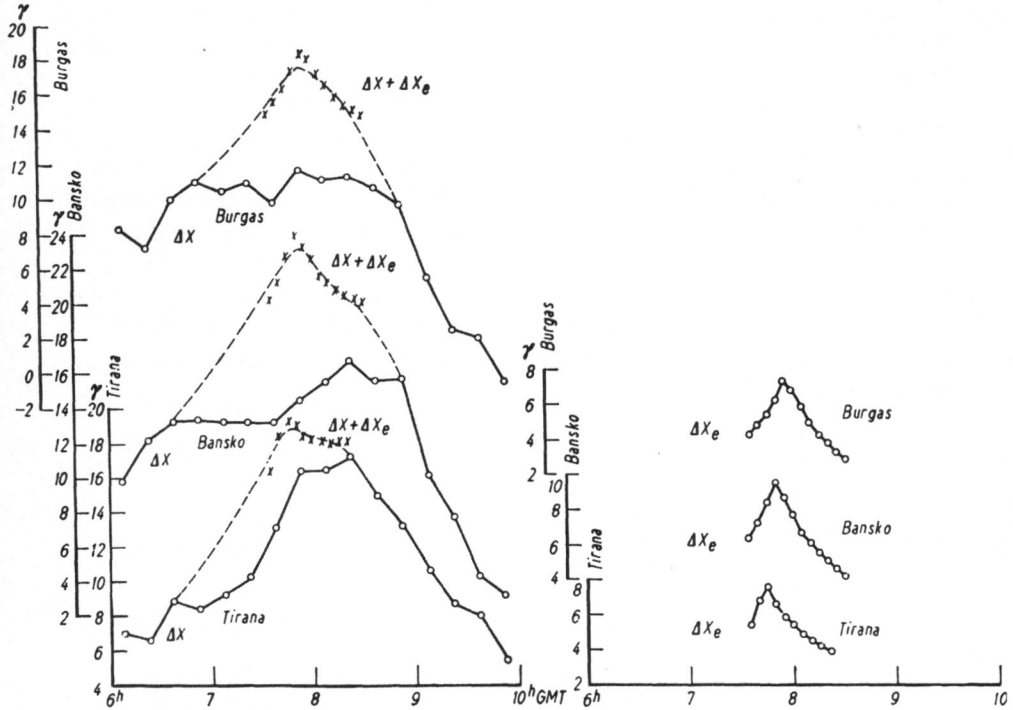

Fig. 8.9. North component ΔX of geomagnetic variation during a solar eclipse on February 15, 1961 (*solid lines, left*), and extrapolated *Sq* variation (*dashed lines, left*) at three stations in South-eastern Europe. *Curves on the right* give the solar eclipse effect as the difference between dashed and solid lines on the left. (Wagner 1963) (1 γ = 1 nT)

$$\sigma U \simeq -\sigma_1 U_1/2 + \sigma_0 U_1 \cos \tau - (\sigma_1 U_1/2) \cos 2\tau \ldots \qquad (8.5)$$

Hence, with the reasonable assumption of $\sigma_1 \simeq \sigma_0$, the relationship in Fig. 8.4 between the two first Fourier components in amplitude ($h_2 \simeq h_1/2$) and phase ($\gamma_2 \simeq \gamma_1 + \pi$) is verified.

Solar control of *Sq* can be detected in the course of the 11-year solar cycle. The amplitude of *Sq* increases by a factor of more than 2 from sunspot minimum to sunspot maximum (Campbell and Matsushita 1982). Two thirds of this increase may result from the enhancement of the electric conductivity with solar activity. The rest is probably due to the increase of the tidal wind speed at dynamo region heights caused by the temperature increase with increasing solar activity.

Solar eclipses cause the electron density and thus the electric conductivity to decrease within the shadow zone of about 6000 km diameter. This electric "hole" moves with the shadow zone and locally diminishes the strength of the *Sq* current. The magnetic effect on the ground is a small reduction of the horizontal component which can only be detected during geomagnetically quiet conditions. Figure 8.9 shows the north component ΔX of the geomagnetic variation during the solar eclipse of February 15, 1961, and the extrapolated *Sq* variation (dashed lines) at stations in south eastern Europe. The curves on the right hand side give the residual geomagnetic eclipse effect which is of the order of 5 to 10 nT.

The opposite effect of an enhancement of *Sq* is observed during a solar flare. A solar flare produces a burst of solar radiation from the environment of an active sunspot. The XUV component of the burst increases the electron density in the sunlit hemisphere mainly within the ionospheric *D* and *E* region for an interval of about 1 h. The electric conductivity increases, and the *Sq* current is enhanced. The magnetic effect on the ground − a small deviation from the undisturbed *Sq* pattern − is called a crochet or a geomagnetic solar flare effect. The corresponding electric current is similar to that of the undisturbed *Sq* current, except shifted in phase by a few degrees (Richmond and Venkateswaran 1971).

9 Ionospheric Dynamo

We develop in this chapter a more quantitative treatment of the ionospheric dynamo. For this reason, it is necessary to recall some basic knowledge about atmospheric tidal theory and the coupling between tidal winds and the ionospheric plasma.

9.1 Elements of Tidal Theory

The atmosphere behaves like a wave guide for pressure and wind fluctuations, closed on the ground and open to space (Fig. 9.1). Only individual wave modes can exist in such a wave guide. They are characterized by a separation constant h_n^m, called the equivalent depth. The wave structure of the pressure of a tidal wave is given (e.g., Chapman and Lindzen 1970) by

$$p/p_0 = p_n^m(z)\, \Theta_n^m(\theta)\, \exp\left[im(\lambda + \Omega_i t)\right] , \qquad (9.1)$$

where p_0 is the background pressure, and (z, θ, λ) are height, colatitude, and longitude (Fig. 14.1d). The meridional structure is described by the Hough function Θ_n^m, m is the zonal wave number, and $m\,\Omega_i$ is the angular frequency of the westward migrating tidal waves. $i = S$ corresponds to a period of one solar day, $i = L$ corresponds to the period of one lunar day, so that the tidal waves depend on local (solar or lunar) time:

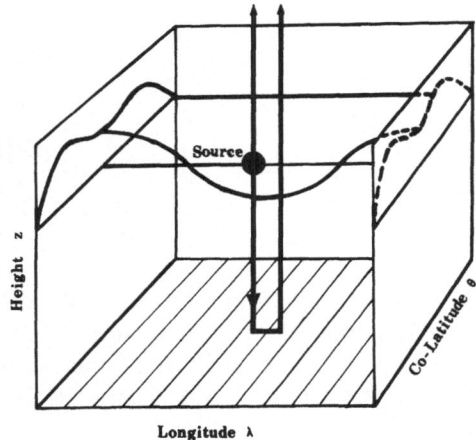

Fig. 9.1. Schematic view of atmospheric wave guide in which the diurnal tidal wave (1, −2) propagates. *Dashed curve* gives its meridional structure, *solid curve* gives its zonal structure

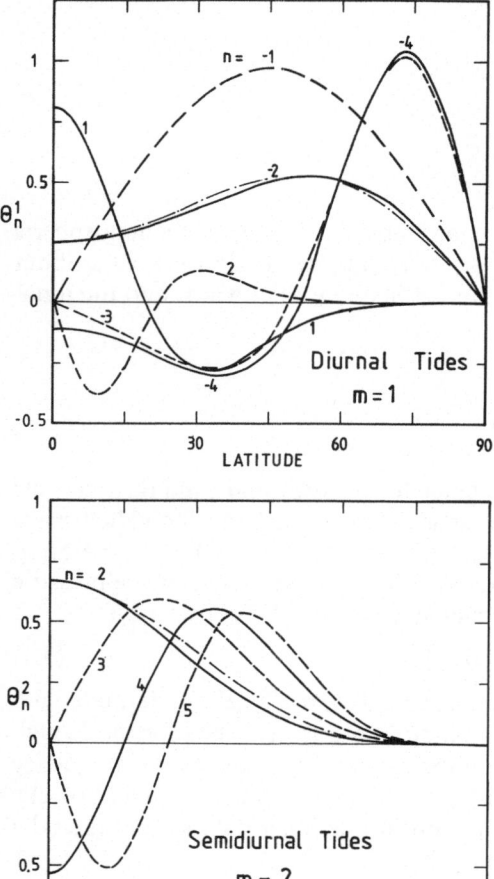

Fig. 9.2. Hough functions of diurnal tides versus latitude describing pressure variations of the tidal wave modes. Symmetric modes (with respect to the equator) are plotted as *solid curves*. Antisymmetric waves are plotted as *dashed curves*. *Dash-dotted curve* is an approximation of $(1, -2)$ from Eq. (9.13)

Fig. 9.3. Hough functions of semidiurnal tides. Otherwise as in Fig. 9.2. *Dash-dotted curve* is an approximation of $(2, 2)$ from Eq. (9.20)

$$\tau_i = \lambda + \Omega_i t . \tag{9.2}$$

These quantities are related to the frequency of one siderial day Ω: $\Omega_S/\Omega = 0.9973$ and $\Omega_L/\Omega = 0.9635$ with $\Omega = 7.292 \times 10^{-5}\,\text{s}^{-1}$. For our purpose, it is reasonable to set $\Omega_i/\Omega \simeq 1$.

Figure 9.2 shows the Hough functions of some important diurnal tides ($m = 1$). Figure 9.3 shows the Hough functions of some semidiurnal tides ($m = 2$). The wave modes are identified according to their zonal and meridional wave numbers (m, n). Two classes of wave exist: waves of class I defined by positive numbers of n, and waves of class II with negative n.

The height structure function $p_n^m(z)$ for the case of a free wave, ascending in an isothermal, nondissipative atmosphere, is given by

Table 9.1. Equivalent depths of various solar diurnal and semidiurnal tidal wave modes on the ground. (Adapted from Chapman and Lindzen 1970)

Wave mode	Equivalent depth h (km)	
$(1, -1)$	803.356	Diurnal
$(1, -2)$	-12.270	
$(1, -3)$	-1.814	
$(1, -4)$	-1.758	
$(1, 1)$	0.691	
$(1, 2)$	0.238	
$(2, 2)$	7.852	Semidiurnal
$(2, 3)$	3.667	
$(2, 4)$	2.110	
$(2, 5)$	1.367	

$$p_n^m(z) = \bar{p} \exp[(1 - i\alpha)z/(2H)] \tag{9.3}$$

with

$$\alpha = [4\kappa H/h_n^m - 1]^{1/2} \tag{9.4}$$

where H is the pressure scale height, $\kappa = (c_p - c_v)/c_p$, with c_p, c_v the specific heats at constant pressure and volume, respectively, and h_n^m is the equivalent depth.

Table 9.1 gives the values of the equivalent depths for the tidal waves from Figs. 9.2 and 9.3. Waves of the first class have $h_n^m > 0$. Most tidal waves of the second class have $h_n^m < 0$. The numbering of n is according to the sequence of decreasing h. Waves with real α, for which $4\kappa H/h > 1$, are called internal waves. Their vertical wave lengths are

$$\lambda_z = 4\pi H/\alpha. \tag{9.5}$$

If $4\kappa H/h < 1$, it is $\alpha = i\beta$ with β real. These waves are called external waves. Their amplitudes decrease according to $\exp(-z/H_z)$ where

$$H_z = 2H/(\beta - 1) \tag{9.6}$$

is an attenuation scale height. The external waves do not have a vertical wave structure. Evidently, the diurnal $(1, -2)$ wave is an external wave.

The tidal waves are excited either by solar heat input or by the lunar gravitational force. Therefore, the height structure functions of forced tides become rather complex and can, in general, only be determined by numerical methods. Moreover, the tides are modified by the variable zonal background wind, mainly at middle atmospheric heights (Forbes 1982a, b).

The tidal waves are dissipated by molecular viscosity, heat conduction and ion drag at ionospheric heights. This wave dissipation can be described by a complex equivalent depth, so that the vertical structure is now determined by $\exp\langle[(1 - \beta) - i\alpha]z/(2H)\rangle$ with (α, β) real. The difference between class I and

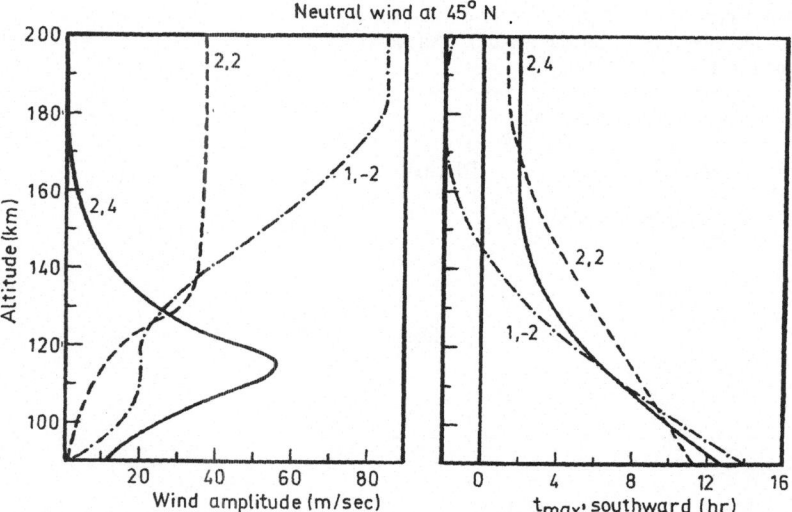

Fig. 9.4. Height profiles of southward wind components of the three dominant solar tides in amplitude (*left*), and time of maximum (*right*) within the dynamo region at 45° latitude. (Salah and Evans 1977)

class II waves disappears, and all tidal waves eventually become quasi-external waves with $\beta > 1$ at heights above about 150 km (e.g., Volland and Mayr 1977).

Figure 9.4 shows observed winds decomposed into their solar tidal components within the dynamo region. The dominance of the diurnal $(1, -2)$ mode above 140 km altitude is evident.

9.2 Coupling Between Tidal Wind and Ionospheric Plasma

We analyze in this section the dynamo action of the tidal wind at ionospheric heights. This coupling can be described by the horizontal momentum equation of the neutral wind in a rotating atmosphere together with an equation for the divergence of the horizontal wind:

$$\partial U/\partial t + 2\Omega \hat{r} \times U \cos\theta + (1/\varrho_0)\nabla p = J \times B_0/(\varrho_0 \Delta h) \tag{9.7}$$

$$\nabla \cdot U + 1/(\varrho_0 gh)\partial p/\partial t = 0 .$$

The momentum equation balances the inertial force, the Coriolis force, the horizontal pressure gradient and the Ampere force. $\hat{r}$ is the unit vector in radial direction. ∇ is the horizontal nabla operator in spherical coordinates. The Ampere force $J \times B_0$ couples the electric sheet current J to the horizontal wind U (Fig. 9.5). ϱ_0 is the background density in the center of the dynamo region. The equivalent depth h determines the divergence of the horizontal wind.

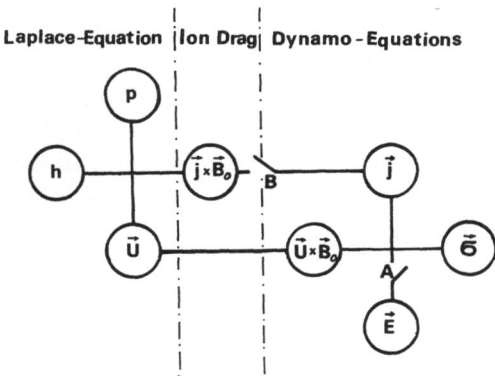

Fig. 9.5. Block diagram illustrating coupling between horizontal wind U, pressure p, electric current density j, and electric field E via Ampere force $j \times B_0$ and Lorentz force $U \times B_0$. B_0 is the geomagnetic field, h is the equivalent depth, and σ is the conductivity tensor. In a self-consistent treatment of the coupled system of Eqs. (9.7) and (9.8), gate B must be closed. In conventional dynamo theories, gate B is open

The electric current must obey Ohm's law and the condition that no sources or sinks exist:

$$J = \Sigma \cdot (E + U \times B_0) \tag{9.8}$$
$$\nabla \cdot J = 0 .$$

Here, we have already integrated over the dynamo region of thickness Δh so that U is averaged over this area, and Σ is the height-integrated conductivity tensor from Eq. (2.20). It is obvious that only those wind systems contribute significantly to the height-integrated current J that have vertical wavelengths large compared to the thickness of the dynamo layer. The feedback between wind and plasma in Eq. (9.8) is via the Lorentz field $U \times B_0$ (Fig. 9.5). The electric field E is a secondary polarization field set up by the charge separation between the ions and the electrons. It adjusts the electric current so that the source free condition for J is fulfilled.

For an approximate solution of the set of Eq. (9.7) and (9.8), we first assume a coaxial geomagnetic dipole field with its field components according to Eq. (14.3).

Furthermore, the electric polarization field is considered as a potential field:

$$E = - \nabla \Phi . \tag{9.9}$$

Finally, the elements of the conductivity tensor in Eq. (2.20) are taken as constant, and the current is assumed to flow in a horizontal sheet. Separating the horizontal component of $U \times B_0$ into a curl-free and a source-free component:

$$U \times B_0 = - \nabla V + \nabla \times (W \hat{r}) , \tag{9.10}$$

where V and W are functions of (θ, λ) only, we find from Eq. (9.8) (Moehlmann 1974)

$$\Phi = - V - (\Sigma_h / \Sigma_p) W , \tag{9.11}$$

which indicates that the curl-free component V of the Lorentz field is partly compensated by the electrostatic potential Φ. Furthermore,

$$J = \nabla \times (\Psi \hat{r}), \tag{9.12}$$

where $\Psi = \Sigma_c W$ is a stream function and $\Sigma_c = \Sigma_p + \Sigma_h^2/\Sigma_p$ is a height-integrated Cowling conductivity. The lines $\Psi = $ constant give the stream lines of the current, the current flowing to the right orthogonal to the direction of increasing Ψ. Only the source-free component of the Lorentz force drives the electric current. The effective conductivity has increased because the current is forced to remain in the horizontal thin sheet.

9.3 Diurnal (1, −2) Mode

An approximate solution of the fundamental solar diurnal (1, −2) mode is possible if one introduces into Eqs. (9.7), (9.10) and (9.12) the expressions (Volland 1976)

$$\left.\begin{aligned}
p &= (iU_0 a \Omega \varrho_0/5)(1 + 4\cos^2\theta)\sin\theta \\
u &= U_0 b \cos\theta \\
v &= -(iU_0/5)(1 - 6s\cos^2\theta) \\
V &= (U_0 B_{00} ai/5)(4s - 1 + 4s\cos^2\theta)\sin\theta \\
W &= -(U_0 B_{00} a/5)(4s - 1)\sin\theta\cos\theta
\end{aligned}\right\} \exp(i\tau) \tag{9.13}$$

and

$$h = -9a^2\Omega^2 d/(20g). \tag{9.14}$$

The two momentum equations in (9.7) are exactly fulfilled and the second equation in (9.7) is approximately fulfilled if the following numbers are introduced into Eq. (9.13):

$$\begin{aligned}
b &= (1 + 2i\delta/15)/(1 + 4i\delta/3) \\
s &= (1 + i\delta/3)/(1 + 4i\delta/3) \\
d &= (1 + 4i\delta/3)/(1 - i\delta/6),
\end{aligned} \tag{9.15}$$

with $\delta = \Sigma_c B_{00}^2/(\Delta h \Omega \varrho_0)$.

In the case of no plasma interaction, δ goes to zero, and $h = -9.9$ km. This roughly approximates the exact number of -12.3 km of the (1, −2) mode in Table 9.1. The dash-dotted line in Fig. 9.2 gives the meridional structure of p from Eq. (9.13) and indicates the degree of approximation of the Hough function (1, −2).

Figure 9.6 shows the equipotential lines of the electric field, and Fig. 9.7 gives the streamlines of the electric current, determined from Eq. (9.14) for the cases $\delta = 0$ (no feedback between plasma and wind; model 1), and $\delta = 1$ (feedback between plasma and wind; model 2), and with the numbers $\Sigma_c = 60$ S; $\Sigma_h/\Sigma_p = 2$; $\Delta h = 50$ km; $U_0 = 40$ m/s; $\varrho_0 = 1.5 \times 10^{-8}$ kg/m^3 (at 125 km height).

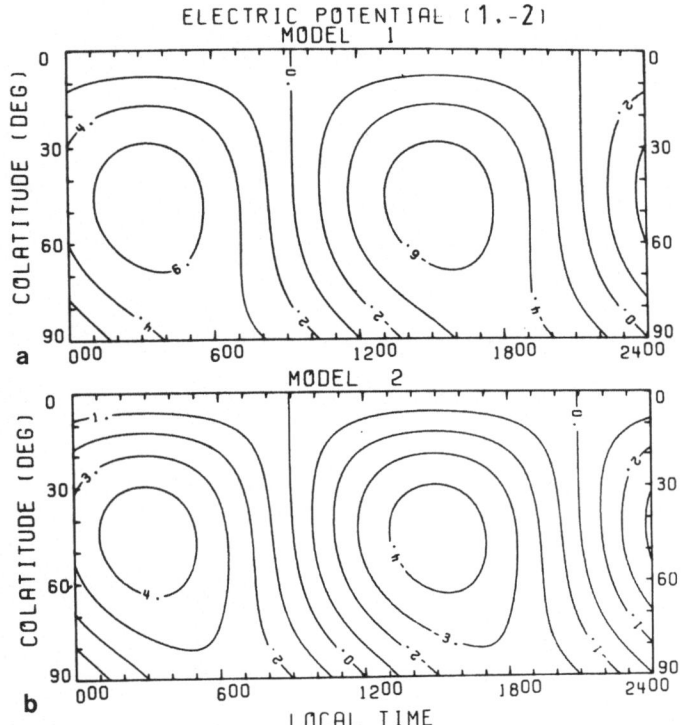

Fig. 9.6a,b. Equipotential lines of electric polarization potential of solar diurnal $(1, -2)$ tidal mode for $\delta = 0$ (model 1; upper panel) and $\delta = 1$ (model 2; lower panel) determined from Eqs. (9.11) and (9.12). Between two lines is a voltage difference of 1 kV and 2 kV, respectively

Note that the Hall conductivity Σ_h changes sign in the southern hemisphere, and W in Eq. (9.10) is antisymmetric with respect to the equator. The electric potential Φ from Eq. (9.11) is therefore symmetric to the equator. The value of Φ is not realistic at the equator where one expects $E_\theta = 0$.

The main effect of the feedback in Figs. 9.6b and 9.7b is that field and current decrease in amplitude by about 1/3 compared with the numbers in Figs. 9.6a and 9.7a, although the pattern remains similar with and without feedback. The pattern of the electric potential in Fig. 9.6b is similar to the diurnal component of the observed Sq field in Fig. 8.8. The Sq current in Fig. 9.6 is, of course, only the first Fourier component of the real current in Fig. 8.3a and has the right magnitude and the right pattern of this first Fourier component.

The total electric current within one vortex is given by

$$\Psi_m = |\Psi|_{\theta = \pi/4} = 3 |U_0| B_{00} a \Sigma_c / [10(1 + 16\delta^2/9)]^{1/2} \tag{9.16}$$

and becomes $\Psi_m = 83$ kA for $\delta = 1$ and $\Psi_m = 138$ kA for $\delta = 0$. The efficiency of the diurnal tide in producing the dynamo current is, therefore, only about 60% of that of the open circuit ($\delta = 0$).

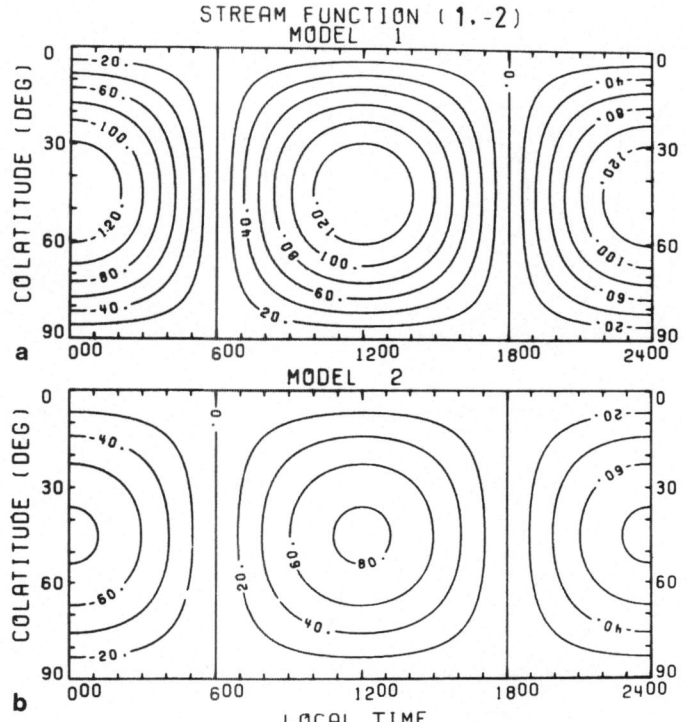

Fig. 9.7a, b. Stream lines of electric current of solar diurnal $(1, -2)$ tidal mode. Otherwise as in Fig. 9.6. Between two stream lines flows a current of 20 kA

The short circuit current $(\delta, \Sigma_c \to \infty)$ becomes

$$\Psi_m \to 9 \Delta h \Omega \varrho_0 |U_0| a/(40 B_{00}) = 105 \text{ kA} ,\tag{9.17}$$

which means that the current is relatively insensitive to the height-integrated conductivity.

The maximum electric potential is much more sensitive to the electric conductivity. For $\delta \to \infty$, one has

$$\Phi_m \to V_m \to 2 |U_0| B_{00} a/(15 \sqrt{3}) = 590 \text{ V} ,\tag{9.18}$$

which is smaller by a factor of 7 than the maximum potential in Fig. 9.6b.

We can estimate the Joule heating produced by the $(1, -2)$ current averaged over the sphere. This is

$$P_{\text{joule}} = 1/(2\Sigma_c) \int_0^\pi \text{Real} (J \cdot J^*) \sin \theta \, d\theta = 6 U_0^2 B_{00}^2 \Sigma_c /\{[125(1 + 16 \delta^2/9)]\}\tag{9.19}$$

which gives 1.5 μW/m² for $\delta = 1$. This is more than two orders of magnitude smaller than the solar XUV heat input into the upper atmosphere above 120 km height and is therefore of minor importance.

9.4 Semidiurnal (2, 2) Mode

In a manner similar as for the diurnal $(1, -2)$ mode, one finds approximate solutions for the semidiurnal (solar or lunar) $(2, 2)$ mode. These are (Volland 1976)

$$
\left.
\begin{aligned}
p &= -(2U_0 a \Omega \varrho_0/3)(1 - f\cos^2\theta)\sin^2\theta \\
u &= -2iU_0 b \sin\theta\cos\theta \\
v &= (2U_0/3)(1 + 2s\cos^2\theta)\sin\theta \\
V &= (2U_0 B_{00} a/9)(1 + s + 3s\cos^2\theta)\sin^2\theta \\
W &= (4iU_0 B_{00} a/9)(1 + s)\sin^2\theta\cos\theta
\end{aligned}
\right\} \exp(2i\tau) \qquad (9.20)
$$

$$
h = 12a^2\Omega^2 d/(35\,\mathrm{g})
$$

with

$$
\begin{aligned}
f &= (1 + 7i\delta/3)/(1 - i\delta/3) \\
b &= (1 + i\delta/9)/(1 - i\delta/3) \\
s &= (1 + i\delta/3)/(1 - i\delta/3) \\
d &= (1 - 7i\delta/9)/(1 - i\delta/15) \,.
\end{aligned}
$$

The dash-dotted curve in Fig. 9.3 is from Eq. (9.20) and shows the degree of approximation of the (2, 2) mode. The value of h from Eq. (9.20) is 7.5 km, as compared with the exact number of 7.85 km from Table 9.1.

Figure 9.8 shows the stream function of the current of the (2, 2) mode derived from Eq. (9.20) for $\delta = 0$ (model 1) and $\delta = 1$ (model 2), respectively, with $U_0 = 0.68$ m/s and the other numbers as in Fig. 9.7. The shape of the stream function in Fig. 9.8 is in reasonable agreement with the shape of the observed L current in Fig. 8.5. It would also agree in magnitude if the wind amplitude were $U_0 = 1.2$ m/s.

The total current within one vortex in Fig. 9.8 is given by

$$
\Psi_m = 16\Sigma_c |U_0| B_{00} a/[27(3 + \delta^2/3)^{1/2}] \simeq 2.5\,\mathrm{kA}\,. \qquad (9.21)
$$

By comparison, the maximum short circuit current $(\delta = \infty)$ is 8.1 kA, and the maximum open circuit current $(\delta = 0)$ is 2.7 kA.

The Joule heating of the lunar (2, 2) mode averaged over the sphere becomes

$$
P_{\mathrm{joule}} = 0.06\,\Sigma_c U_0^2 B_{00}^2/(1 + \delta^2/9) \qquad (9.22)
$$

which gives $4\,\mathrm{nW/m^2}$ for $\delta = 1$ and $U_0 = 1.2$ m/s. This is more than two orders of magnitude smaller than the corresponding value for the $(1, -2)$ mode in Eq. (9.19).

Numerical approaches to determine J and E usually start from calculated tidal winds U and use these theoretical winds as "driving force" to evaluate the electric fields and currents from Eq. (9.8) (Stening 1973; Richmond et al.

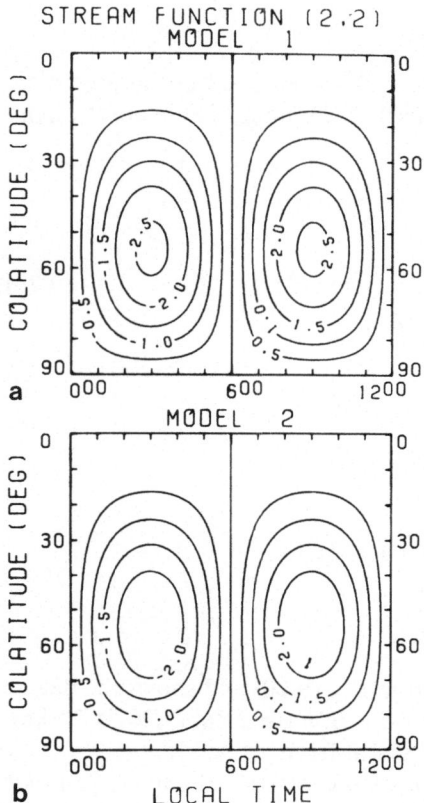

Fig. 9.8a, b. Stream lines of electric current of lunar semidiurnal (2, 2) tidal mode determined from Eqs. (9.12) and (9.20). **a** Model 1 ($\delta = 0$); **b** model 2 ($\delta = 1$). Between two stream lines flows a current of 0.5 kA

1976; Forbes and Lindzen 1976a, b; Moehlmann 1977; Takeda and Maeda 1980, 1981). An inverse method to determine wind and electric field from the observed Sq current was applied by Kato (1956).

The principle results of these numerical studies are the following: The diurnal $(1, -2)$ mode contributes about 70 to 80% of the total Sq current. The balance of 20 to 30% comes from the diurnal $(1, 1)$ mode and the semidiurnal $(2, 2)$ and $(2, 4)$ modes. These internal modes with their finite vertical wavelengths are mainly responsible for the observed variability in Sq. A smaller part of the Sq current may flow outside the dynamo region along the geomagnetic fieldlines within the magnetosphere, especially during solstice conditions where summer and winter hemisphere have a small electric potential difference (Maeda 1974; Wagner et al. 1980). The bulk of the Pedersen current flows in a broad region between about 120 and 170 km altitude, whereas the Hall current is concentrated between 100 and 120 km height. The contribution of the Hall current is of the same order of magnitude as that of the Pedersen current, yielding $\Sigma_h/\Sigma_p \simeq 1$ and $\Sigma_c \simeq 2\,\Sigma_p$. The lunar tides are predominantly excited by the lunar semidiurnal $(2, 2)$ mode (Tarpley 1970). A review of dynamo theories is given by Kato (1980).

9.5 Equivalent Electric Circuit of *Sq*

The feedback between tidal winds and ionospheric plasma is schematically illustrated in Fig. 9.5. The various physical parameters from Eqs. (9.7) and (9.8) are connected in this block diagram. As already mentioned, coupling from the wind to the plasma occurs via the electric Lorentz field $U \times B_0$ in Ohm's law [Eq. (9.8)], whereas feedback from the plasma to the wind is via the mechanical Ampere force $J \times B_0$ in the momentum equation (9.7). Gate *B* in Fig. 9.5 must be closed in order to account for this mutual feedback consistently, as outlined in Sections 9.3 and 9.4. Such a dynamo is called a hydromagnetic dynamo.

In conventional dynamo theories (e.g., Stening 1969; Tarpley 1970), gate *B* is open, so that the wind is considered as an independent "external driving force". Such a dynamo is called a kinematic dynamo. In our treatment in Sections 9.3 and 9.4, this condition corresponds to $\delta = 0$. Clearly, kinematic dynamos overestimate the efficiency of the wind to excite electric currents. In particular, they become increasingly unrealistic for increasing values of δ and cannot treat the case of a short circuit ($\delta \to \infty$).

More sophisticated dynamo theories proceed in two steps. First, gate *B* is closed and gate *A* is left open, so that the wind structure is calculated without taking the electric field into account. This means that the Ampere force in Eq. (9.7) is replaced by $\Sigma \cdot (U \times B_0) \times B_0 \simeq -\nu \cdot U$, where ν is called an ion drag tensor (Murata 1974). In the simplest case where only the Pedersen conductivity is taken into account, ν becomes a scalar. The second step is to introduce winds, so modified by ion drag, into the dynamo equation (9.8) (now closing gate *A* and opening gate *B* in Fig. 9.5) in order to calculate the electric current and field (e.g., Richmond et al. 1976).

We want to find an equivalent electric circuit of the *Sq* current in which the importance of gate *B* in Fig. 9.5 becomes evident. We consider winds and currents of the $(1, -2)$ mode integrated along a meridional circle. Then, the meridional components disappear, and the integrated zonal components of the Lorentz potential *U* and the electric current *J* become

$$U = a \int_0^\pi U \times B_0 \sin \theta \, d\theta = b \bar{U}$$

$$\tag{9.23}$$

$$J = a \int_0^\pi J \sin \theta \, d\theta = -(4s-1) \bar{U}/(10R) \,,$$

with

$$\bar{U} = 4U_0 B_{00} a/3; \quad R = 1/\Sigma_c; \quad \text{and} \quad b, s \text{ from Eq. (9.15)}.$$

We introduce two resistances, R_i and R_p, which obey the conditions

$$U - \bar{U} = -R_i J$$

$$RJ = U_p + U$$

$$\tag{9.24}$$

$$U_p = -R_p J \,,$$

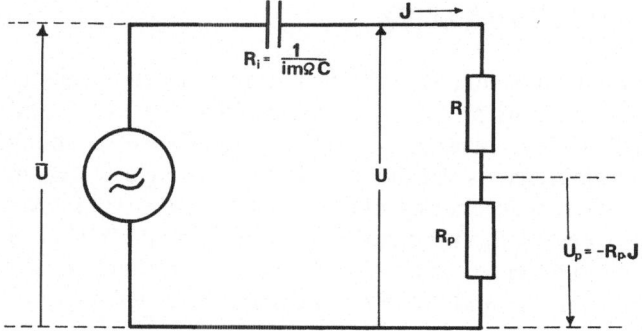

Fig. 9.9. Equivalent electric circuit of the ionospheric dynamo. The source voltage $\bar{U}$ corresponds to the driving solar heat input. The ionospheric dynamo has an internal resistance R_i which behaves like a capacitance C. J is the electric dynamo current, the load resistance R corresponds to the electric conductivity of the dynamo region. The voltage U is a measure of the wind velocity. The complex impedance R_p is a measure of the secondary electric polarization voltage U_p

and obtain from Eqs. (9.12), (9.13) and (9.23)

$$R_i = 1/(i\Omega C); \quad R_p = -13R/3 + R_i/9; \quad C = 1/(4\Omega R \delta) .$$

The first equation in (9.24) simulates the momentum equation in (9.7) where U is a measure of the wind, $\bar{U}$ is a measure of the driving pressure, and $R_i J$ is a measure of the Ampere force. The second equation in (9.24) simulates Ohm's law where U_p is the polarization potential and R the load resistance.

Figure 9.9 shows an equivalent electric circuit of the ionospheric dynamo. $\bar{U}$ is here the source voltage, R_i the internal resistance of the voltage source which is independent of the plasma parameters and behaves like a capacitance C, J is the electric current, U the Lorentz voltage, R the load resistance depending on the electric conductivity of the dynamo region, and R_p is a complex impedance which is a measure of charge separation between electrons and ions. In a technical dynamo, $\bar{U}$ is the electric analogy of the mechanical force that rotates the coil in an external magnetic field, R_i is the electric analogy of the mechanical resistance of the rotating coil, and R_p corresponds to the self-impedance of the coil.

Voltage and current in Eq. (9.24) and Fig. 9.9 are related according to

$$U/\bar{U} = (R + R_p)/(R + R_i + R_p)$$

$$J/\bar{U} = 1/(R + R_i + R_p) .$$

(9.25)

The importance of the two impedances R_i and R_p becomes clear, if one considers the two extreme cases of a short circuit ($\delta \to \infty$) and an open circuit ($\delta = 0$). The open circuit has $R_i = 0$, and therefore $U = \bar{U}$ and $J = -3\bar{U}/(10R)$, while the short circuit ($R = 0$) has $U = \bar{U}/10$ and $J = 9\bar{U}/(10R_i)$. During a short circuit, the wind amplitude brakes down to 10% of its open circuit value while the current reaches a maximum, but still finite, value because the self-impedance (charge separation) prevents the current from becoming infinite.

9.6 Equatorial Electrojet

The geomagnetic field at mid and high latitudes is predominantly vertically orientated. The tides have primarily horizontal wind directions and therefore induce horizontally directed electric fields that move the plasma in a horizontal sheet.

In contrast to this, the geomagnetic field vector is horizontal and northward-directed at the magnetic dip equator. During equinox conditions, the electric polarization field of the Sq current and the tidal winds have primarily east−west components at the equator. The zonal electric field generates vertical Hall currents which are, however, not divergence-free and are therefore connected with meridional currents. For reasons of symmetry, the meridional current must disappear at the equator during equinox. Moreover, polarization charges will be built up in the vertical direction due to the limited vertical extent of the conducting dynamo region. These polarization charges produce a vertical electric field at the equator. A zonal Hall current driven by this secondary electric field is superposed on the original zonal Pedersen current (see Fig. 2.5). The total current strength in the immediate environment of the equator is thus enhanced. This enhancement can be described by a height integrated Cowling conductivity [see Eq. (2.18)]:

$$\hat{\Sigma}_c = \int_{h_1}^{h_2} \sigma_c dz \,. \tag{9.26}$$

This integrating Cowling conductivity is a factor of about four greater than the conductivity Σ_c at mid and high latitudes as defined in Eq. (9.12).

This strong enhancement of the conductivity near the equator is the reason for the existence of a band of eastward flowing currents between the morning and early afternoon hours − the equatorial electrojet (e.g., Onwumechilli 1967; Forbes 1981). One can estimate the meridional range of the electrojet by considering the formula of the electric conductivity in Eq. (2.19). The contrast between the "equatorial behavior" and the "midlatitude behavior" is essentially given by the dominance of the first or the second term of D in Eq. (2.19). If $\sigma_F \sin^2 I < \sigma_p \cos^2 I$, the "equatorial behavior" is valid. The transition occurs at

$$I = \arctan\left[(\sigma_p/\sigma_F)^{1/2}\right] \simeq 2\phi \simeq 2.9° \tag{9.27}$$

[with $\sigma_F/\sigma_p \simeq 400$, and the relation between I and ϕ from Eq. (14.7)]. This limits the range where the equatorial electrojet flows. It corresponds to a distance of about 160 km from the equator.

Solar as well as lunar tides contribute to the equatorial electrojet. Figure 9.10 shows a succession of 8 days of the horizontal intensity ΔH at Huancayo, a station in Peru located at the dip equator, during quiet conditions. The curves of the L variation and the combination of $L + Sq$ are separated for December solstice (upper panel) and for June solstice (lower panel).

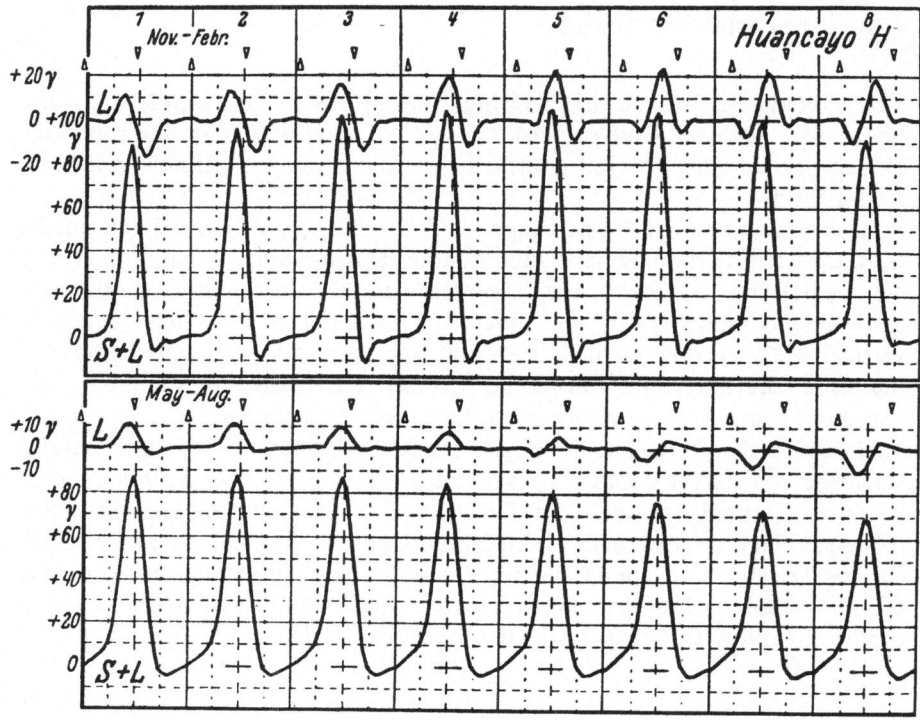

Fig. 9.10. Lunar (L) and combined solar and lunar ($S+L$) daily variations of the horizontal intensity ΔH during quiet days measured at Huancyao, Peru, during November—February (*upper panel*) and May—August (*lower panel*), 1922–1924. (Chapman and Bartels 1951). (1 γ = 1 nT)

The variability of the solar tides gives rise to an occasional reversal of the electrojet in the afternoon. This reversal is called a counter electrojet. During such an event, the semidiurnal solar and lunar tides dominate (Hanuise et al. 1983). The equatorial sporadic E layer disappears and the east—west electric field reverses (Woodman et al. 1977).

We now make a simple calculation of the meridional current system, which develops with the regular equatorial electrojet. We apply a two-dimensional steady-state model and assume a constant and homogeneous electric field E_x, directed to the east. This field is generated by the dynamo action of the tides and drives the meridional secondary currents at the equator. The geomagnetic field is also constant and homogeneous and has only a horizontal component B_y directed to the north. Furthermore, the electric conductivities are assumed to be constant within the height range (h_1, h_2) and zero outside this range. Using Eq. (2.17), one can derive the following relationship between fields and currents (we apply here the coordinate system of Fig. 14.1a):

$$j_x = \sigma_p E_x + \sigma_h E_z = \sigma_c E_x + (\sigma_h/\sigma_p)j_z;$$
$$j_y = \sigma_F E_y; \quad j_z = -\sigma_h E_x + \sigma_p E_z. \tag{9.28}$$

Since current and field depend only on y and z, the current density in the meridional plane can be derived from a stream function:

$$\boldsymbol{j} = (j_y, j_z) = \nabla \times (\Psi \hat{\boldsymbol{x}}) \tag{9.29}$$

or

$$j_y = \partial \Psi / \partial z; \quad j_z = -\partial \Psi / \partial y.$$

Equation (9.29) fulfills the condition that the current is source-free. We assume that the electric field is a potential field from which follows

$$\partial E_z / \partial y = \partial E_y / \partial z. \tag{9.30}$$

The range where the electrojet flows is $(-b < y < b)$, $(h_1 < z < h_2)$. $y = 0$ is the location of the equator, and $h_1 \simeq 90$ km is the lower border of the dynamo region. The boundary conditions are $j_y = 0$ at $y = 0$ and $\pm b$, and $j_z = 0$ at $z = h_1$ and h_2.

After elimination of E_y and E_z from Eqs. (9.28), (9.29) and (9.30), one obtains

$$\partial^2 \Psi / \partial y^2 + \partial^2 \Psi / \partial \bar{z}^2 = 0, \tag{9.31}$$

with $\bar{z} = (\sigma_F / \sigma_p)^{1/2} (z - h_1) \simeq 20(z - h_1)$.

The simplest solution of Eq. (9.31) is a sum of logarithmic potentials which yields, taking into account the boundary conditions: $b, h_2 \to \infty$,

$$\Psi = (A/2) \ln[(y - b_0)^2 + (\bar{z} - \bar{h}_0)^2] + \ln[(y + b_0)^2 + (\bar{z} + \bar{h}_0)^2]$$
$$- \ln[(y - b_0)^2 + (\bar{z} + \bar{h}_0)^2] - \ln[(y + b_0)^2 + (\bar{z} - \bar{h}_0)^2]. \tag{9.32}$$

$(b_0, \bar{h}_0)$ are the coordinates of the center of the meridional-vertical current system. This center is a singular point in Eq. (9.32) where Ψ becomes infinite. Our solution has no physical meaning at this point.

Figure 9.11 shows the stream lines $\Psi = $ const in the northern hemisphere calculated from Eq. (9.32) with $A = 0.1$, $b_0 = 300$ km, $h_0 - h_1 = 30$ km, and $\bar{h}_0 = 600$ km. A total current of 0.3 A/m flows per unit distance in longitude within one vortex.

The stream lines are also isolines of the strength of a toroidal magnetic field $B_x = \mu \Psi$ directed to the west in the northern hemisphere and to the east in the southern hemisphere. One therefore expects a geomagnetic deflection of the declination ΔD of the order of 100 nT near 120 km height decreasing to about 10 nT near 200 km with maximum amplitudes about 1000 km off the dip equator at higher altitudes. The magnetic effect of this current system cannot be detected on the ground. Such magnetic disturbances have indeed been observed from MAGSAT data at heights above 300 km with no counterpart on the ground (Maeda et al. 1982; Takeda and Maeda 1983).

Figure 9.12b shows the vertical electric current density j_z at the equator ($y = 0$) and at $y = 500$ km. Figure 9.12a gives the meridional current density j_y near the center of the vortex ($y = 250$ km) as function of height. These current densities have amplitudes smaller than 10% of the zonal current density

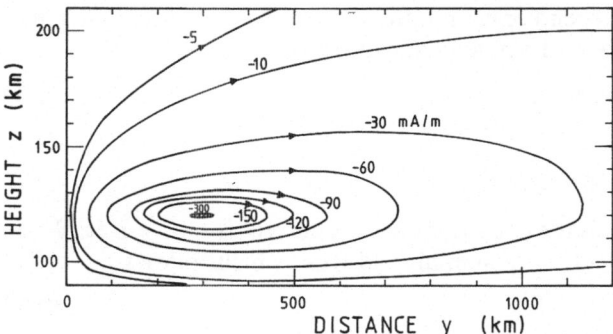

Fig. 9.11. Stream lines of electric meridional current system near the geomagnetic equator as function of northern distance from the equator, and height, as determined from Eq. (9.32). The stream lines are also isolines of a toroidal magnetic field directed to the west. A current system of the same form, but flowing counterclockwise, exists in the southern hemisphere. It is accompanied by a toroidal magnetic field directed to the east

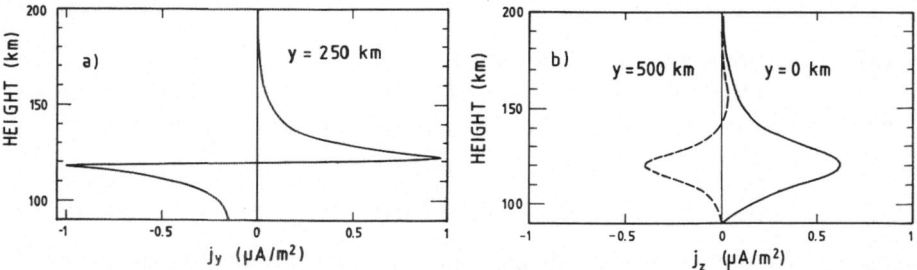

Fig. 9.12. a Meridional current density versus height at 250 km north of the equator, as derived from the equatorial electrojet model in Fig. 9.11. **b** Vertical electric current density versus height at the equator ($y = 0$) (*solid curve*), and at 500 km north of the equator (*dashed curve*)

which is of the order of $j_x \simeq 10\ \mu A/m^2$ (see Fig. 8.7b). However, the vertical electric current density controls j_x, intensifying the electrojet current near the equator and reducing it beyond the vortex where j_z becomes negative [see Eq. (9.28) and Fig. 9.12].

Our simplified model is not self-consistent, as it does not allow a unique determination of center and strength of the vortex of the current system in the meridional plane. More sophisticated models take into account realistic conductivity profiles as well as the dipole configuration of the geomagnetic field (Untiedt 1967; Richmond 1973a, b; Forbes and Lindzen 1976b).

9.7 Upward and Downward Mapping of Electric Sq Field

The dynamo theory assumes the dynamo current to flow in a thin horizontal sheet between about 80 and 200 km altitude. The regions above and below that range are considered as nonconducting. This is approximately true for the middle and lower atmosphere where the conductivity is several orders of magnitude smaller than in the dynamo region. This is also a good assumption for Pedersen and Hall conductivities in the magnetosphere above 200 km. However, it is certainly not true for the parallel conductivity at magnetospheric heights. The large parallel conductivity at these heights prevents the existence of significant parallel electric fields, except possibly in disturbed areas, mainly within the auroral zones (e.g., Stern 1983).

The condition of no parallel electric field along the geomagnetic field lines which is

$$E \cdot B_0 = 0 \tag{9.33}$$

implies that orthogonal fields must map along the geomagnetic field lines. If the electric field is derivable from a quasi-potential Φ, then the equipotential lines of Φ which are orthogonal to E must be parallel to the geomagnetic field lines B_0. In the simplest case of a geomagnetic dipole field [Eq. (14.3)], condition (9.33) is consistent with the condition that Φ is an arbitrary function of the shell parameter L:

$$\Phi = f(L), \tag{9.34}$$

where $L = r/(a \sin^2 \theta)$ is the equation of the dipole lines [Eq. (14.9)].

In the case of the solar tidal $(1, -2)$ mode, the electric potential becomes [see Eqs. (9.11) and (9.13)]

$$\begin{aligned}\Phi = &-(U_0 B_{00} a i/5)[4s - 1 + 4s(1 - 1/L) \\ &+ i(\Sigma_h/\Sigma_p)(4s - 1)(1 - 1/L)^{1/2}](1/L)^{1/2} \exp(i\tau)\end{aligned} \tag{9.35}$$

for $r \geq r_D$ ($r_D \simeq a$ the height of the dynamo region).

The electric field components become

$$E_r = -(\partial \Phi/\partial L)(L/r)$$
$$E_\theta = (\partial \Phi/\partial L)2L \cot \theta/r \tag{9.36}$$
$$E_\lambda = -i\Phi/(r \sin \theta).$$

Figure 9.13 shows the equipotential lines from Eq. (9.35) in the geomagnetic equatorial plane ($\theta = 90°$). Here the same numerical values are applied as in Fig. 9.6. Maeda and Kamei [1975] have estimated the effects of these electric fields on magnetospheric plasma convection where wind velocities of 500 m/s may be reached.

The mapping problem in the realistic nondipole magnetosphere is dealt with by Mozer (1970).

Since the electric conductivity in the middle and lower atmosphere is small but not zero, the electric *Sq* field maps down into these regions. This mapping process can be calculated if a simple exponential law for the conductivity as in Sect. 5.5 is assumed. Then, developing the electric potential in terms of spherical harmonics (in Neumann's normalization) (e.g., Menzel 1960)

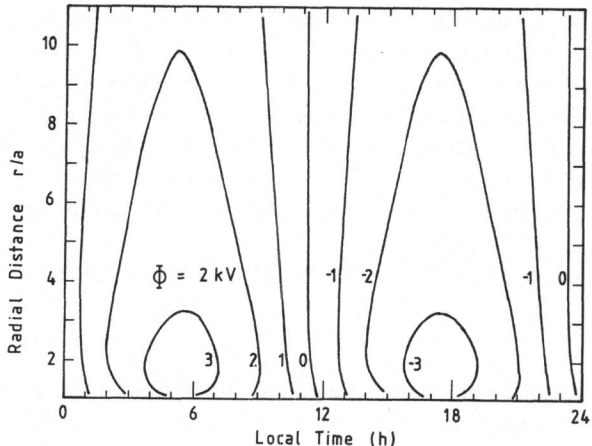

$$\Phi = \sum_n \sum_m \Phi_n^m h_n^m(z) P_n^m(x) \exp(im\tau) \tag{9.37}$$

(with $x = \cos\theta$) one finds the height structure function from Eq. (5.34) which for wave numbers of low degree may be written as

$$h_n^m(z) \simeq h(z) = 1 - \exp(-\alpha z) \tag{9.38}$$

with α the reciprocal scale height of the conductivity from Eq. (5.14). Since $h(z) \simeq 1$ at $z_D = r_D - a \simeq 130$ km, the central height of the dynamo region, the potential of the Sq field in the dynamo region is given by

$$\Phi_D = \Phi(z_D) = \sum \sum \Phi_n^m P_n^m(x) \exp(im\tau). \tag{9.39}$$

For the solar diurnal $(1, -2)$ mode, this gives [see Eqs. (9.11) and (9.13)]

$$\Phi_D = -(U_0 B_{00} ai/5)[(24s/5 - 1)P_1^1(x) + (8s/15)P_3^1(x) + i(4s - 1)/3\,(\Sigma_h/\Sigma_p)P_2^1(x)]\exp(i\tau). \tag{9.40}$$

The electric field components of Eq. (9.37) become

$$\left.\begin{array}{l} E_r = -\alpha\exp(-\alpha z) \\ E_\theta = -h(z)/r \\ E = -h(z)/r \end{array}\right\} \sum\sum \Phi_n^m \left\{\begin{array}{l} P_n^m \\ dP_n^m/d\theta \\ im P_n^m/\sin\theta \end{array}\right\} \exp(im\tau) \tag{9.41}$$

for $a \le r \le r_D$ $(z = r - a)$. Figure 9.14 shows a height–local time cross-section of the equipotential lines of the electric potential [Eq. (9.40)] at 45° latitude. The electric field, which is horizontal at ionospheric heights, bends to the vertical with decreasing height and becomes purely vertical on the ground because the ground is a equipotential plane for quasi-static electric fields. The maximum field strength of the Sq field on the ground is of the order of 1 V/m, which is a factor of about 100 smaller than the thunderstorm field. The horizontal components of the Sq field become smaller than the vertical component below a height where $h(z) \simeq a\alpha\exp(-\alpha z)$. This is near 40 km altitude.

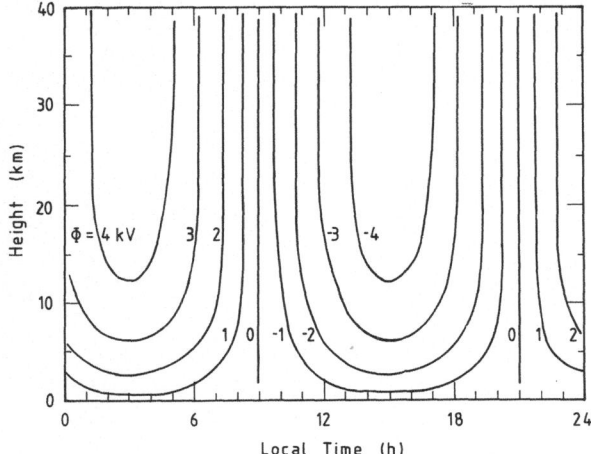

The stream lines of the electric current are roughly orthogonal to the equipotential lines. The middle and lower atmosphere acts like a shunt where part of the *Sq* current flows outside the dynamo region connecting two hemispheres. We can estimate this shunt current by integrating the vertical electric current density which is $j_r = \sigma E_r(z_D)$ over the sphere. For the $(1, -2)$ mode, we arrive at

$$J_r = 4\pi a^2 [(1/2) \int_0^\pi j_r j_r^* \sin\theta \, d\theta]^{1/2} \simeq 36 \text{ A} \qquad (9.42)$$

with $\sigma_0 = 5 \times 10^{-14}$ S/m, $\alpha = 1/6$ km^{-1} and the other numbers from Sect. 9.3. The ratio between the two equivalent resistances in the dynamo region (R_D) and in the lower and middle atmosphere (R_A) is therefore of the order of $R_A/R_D \simeq J_D/J_r \simeq 10^4$, where $J_D = 4\Psi_m = 332$ kA is the total current of the $(1, -2)$ mode (see Fig. 9.7). This ratio is consistent with the ratio R_A/R_D derived from Eqs. (2.8) and (2.21). A more detailed numerical treatment is due to Roble and Hays (1979).

III Solar Wind Interaction with Magnetosphere

10 Coupling Between Solar Wind and Geomagnetic Field

In the next four chapters, we deal with the third main source of electromagnetic energy within the atmosphere: the magnetospheric hydromagnetic dynamo. In this chapter, we present an overview on the phenomenology of the solar wind and its impact on the geomagnetic field, leading to the magnetospheric cavity and to geomagnetic activity. Chapter 11 reviews the global-scale electric field and current configurations within the magnetosphere. In the final two chapters, we discuss briefly low frequency waves and geomagnetic pulsations, which can propagate within the magnetosphere.

10.1 Solar Wind and Interplanetary Magnetic Field

A steady flux of an electrically neutral mixture of low energetic particles, mainly protons and electrons, flows from the sun in a radial direction. This outgasing is called the solar wind. The solar wind velocity varies between about 300 and 900 km/s. The number density of the solar wind particles decreases essentially as the inverse square of the heliocentric distance and reaches values of 10^6 to $10^7\,\mathrm{m}^{-3}$ in the vicinity of the earth (Hundhausen 1979). Magnetic fields are drawn out of the solar surface and are carried by the solar wind into space. The field strength at one AU (the sun−earth distance) is a few nanoteslas.

The general solar magnetic field, to a first approximation an inclined dipole field with magnetic moment $M \simeq 10^{30}\,\mathrm{Am}^2$ and with field strengths on the surface of the order of 0.3 mT, is a quasi-periodic field with a period of about 22 years, so that its dipole moment reverses sign every 11 years. The real solar cycle is, therefore, 22 years rather than 11 years (e.g., Gibson 1973).

The internal solar magnetic dipole field lines are wound up during the course of one half-period of 11 years due to the differential rotation of the surface layers. Strong toroidal (longitudinally directed) magnetic fields develop beneath the visible limb, which eventually break through the surface and cause sunspots − relatively dark areas on the sun with scale sizes of about 10^4 km. The lifetime of sunspots varies from days to months. They often appear in groups of two spots with opposite magnetic polarity (bipolar spots). The overlying magnetic field forms closed loops and can reach field strengths of 0.5 T. The sunspots first appear at heliographic midlatitudes in both hemispheres at the beginning of each cycle, and then drift nearly to the equator during the course of the 11-year sunspot cycle. The number of sunspots reaches a maximum about halfway through the 11-year cycle, coinciding with the time of maximum solar activity. The general solar dipole field reverses shortly after maximum solar activity. The south pole of this dipole field (where the field lines are directed into the sun) has been on the northern hemisphere since 1980 (Hoeksema et al. 1983).

Sunspots tend to be located within certain heliographic zones of longitude. An observer on the earth sees them with a recurrent period of about 27 days – the synodic rotational period of the sun at lower heliographic latitudes. The life time of these active longitudes varies from a few months to a few years.

Solar flares are strong outbursts of visible and XUV radiation occurring usually from the environment of sunspot areas, which last for a few minutes to a few hours. They result from a rapid release of stored magnetic energy that produces plasma heating, particle acceleration, mass ejection and various types of radio emission. Several major solar flares can occur every day during maximum solar activity. Solar flares are moderately rare events during minimum solar activity (sunspot minimum) (e.g., Sturrock 1980).

Regions free of sunspots, in particular the solar polar caps, are magnetically unipolar regions. This means that magnetic field lines of one polarity come out of the solar surface. They are stretched into the interplanetary space by the solar wind (Fig. 10.1). The area of separation between field lines directed away from the sun (away polarity) and those directed toward the sun (toward polarity) is called the interplanetary neutral sheet. This neutral sheet has a wavy configuration like the dress of a dancing ballerina. The earth, which orbits in the ecliptic plane inclined by 7.25° to the heliographic equator, usually passes through two or four sectors of opposite polarity of the interplanetary magnetic field during one synodic rotation of the sun (see Fig. 10.2) depending on the degree of solar activity (Hoeksema et al. 1983).

The unipolar regions on the sun sometimes extend from the polar areas down to low heliographic latitudes and may even cross the equator at certain

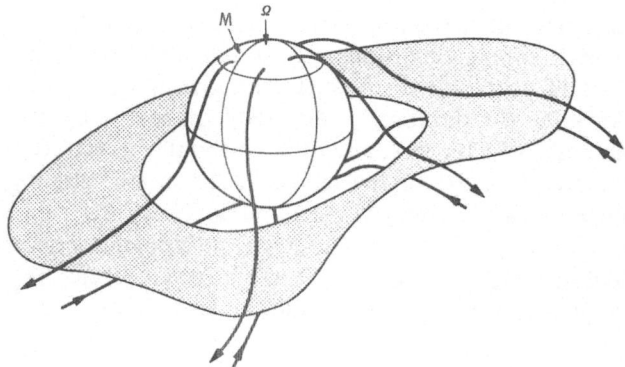

Fig. 10.1. Schematic illustration of magnetic field lines leaving the sun from two unipolar regions. *Hatched area* is a magnetically neutral surface which separates the two regions of unipolar magnetic fields. *M* the inclined axis of the general solar magnetic dipole field; *Ω* the rotational axis. (Smith 1979)

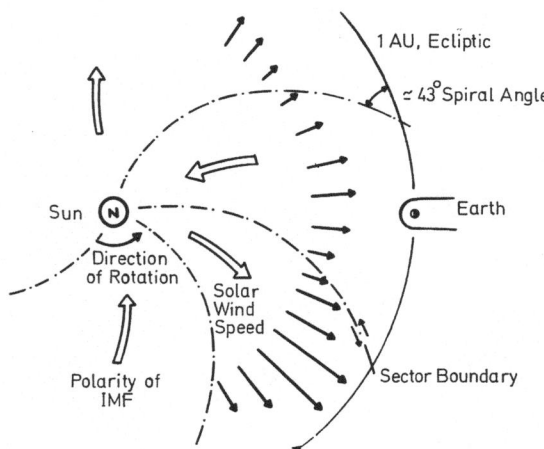

Fig. 10.2. Sector structure of the interplanetary magnetic field (IMF) in the ecliptic plane in the case of four sectors. *Dash-dotted lines* are the sector boundaries. *Open arrows* give the direction of polarity of IMF. *Thin arrows* indicate direction and strength of the solar wind. (Courtesy of G. W. Prölss)

longitudes. These unipolar regions are associated with relatively cold areas of the overlying solar corona — the coronal holes. The solar wind from large coronal holes is accelerated to high speeds ($\simeq 700$ km/s). The average solar wind, however, is only about 400 km/s. Coronal holes which generate high-speed streams of the solar wind that meet the earth's orbit are called M-regions. Coronal holes, as well as the sector structure of the interplanetary magnetic field, rotate almost rigidly with the sun's mean synodical period of about 27.5 days. Their life times range from months to years. Coronal holes tend to be most prominent during the declining phase of the solar cycle (e.g., Zirker 1977).

10.2 Closed Magnetosphere

The plasma of the solar wind impinging on the earth's magnetic field cannot penetrate directly in the earth's atmosphere. It by-passes the geomagnetic field and leaves a cavity — the magnetosphere (Fig. 10.3). The boundary of the magnetosphere — the magnetopause — can be considered as the outermost region of the earth's atmosphere because the ionized particles within the magnetosphere are mainly controlled by the geomagnetic field. Since the solar wind is supersonic, a shock front forms at a distance of several earth radii in front of the subsolar point. The region between the shock front and the magnetopause is called the magnetosheath.

One may idealize the solar wind by a homogeneous, fully ionized, weakly magnetic, unidirectional plasma. The magnetic flux is frozen into such a plasma. If the plasma is essentially field-free, no external magnetic field can penetrate into the plasma. The geomagnetic field surrounded by the solar wind plasma is therefore compressed and confined to a cavity, and electric currents

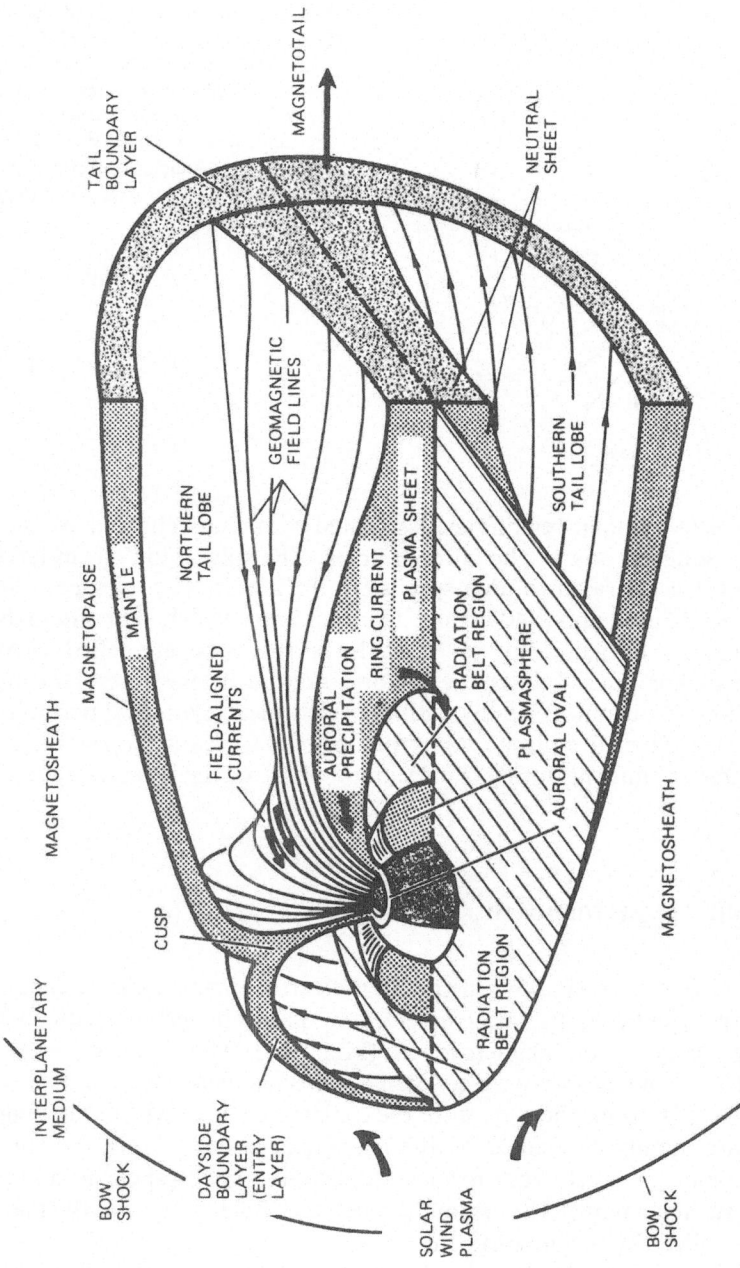

Fig. 10.3. Perspective illustration of magnetosphere. (Courtesy of J. G. Roederer)

must flow on the border of this cavity. They generate magnetic fields which just cancel the magnetic field outside the cavity. Such an idealized cavity is called a closed magnetosphere.

The real magnetosphere behaves to a first approximation like a closed magnetosphere. We estimate the point of closest approach of the solar wind to the center of the earth on the upwind side – the stagnation point or the subsolar point. At this point ($r = r_s$), the solar wind pressure given by $p_s = 2 m_s n_s v_s^2$ (with m_s the mass of the solar wind particles (mainly protons), n_s its number density, and v_s the solar wind velocity), is just cancelled by the magnetic pressure which is $p_m = B_i^2/(2\mu)$. At the stagnation point just inside the magnetopause, the magnetic field B_i is the sum of the dipole field B_0 and the magnetic field B_{cf} due to the compensating boundary currents (the Chapman-Ferraro currents). However, since the magnetic field just outside the boundary must disappear, ($B_0 - B_{cf} = 0$ at $r > r_s$), it follows that inside the boundary it is

$$B_i = 2 B_0 = 2 B_{00} (a/r_s)^3 . \tag{10.1}$$

Thus, the distance of the subsolar point becomes

$$r_s/a = [B_{00}^2/(\mu m_s n_s v_s^2)]^{1/6} \simeq 10 , \tag{10.2}$$

with $n_s \simeq 5 \times 10^6 \, \mathrm{m}^{-3}$, $v_s \simeq 3 \times 10^5 \, \mathrm{m/s}$. This distance usually varies between about 9 and 11 earth radii (Maezawa 1974), but may become as close as 6 earth radii during extremely disturbed conditions.

The magnetic field at the subsolar point is $B_i \simeq 60 \, \mathrm{nT}$. An electric sheet current generating the magnetic field $B_{cf} = 30 \, \mathrm{nT}$ therefore flows from dawn to dusk having a strength of $j_y \Delta x \simeq B_{cf}/(2\mu) \simeq 10^{-2} \, \mathrm{A/m}$ (see Fig. 14.3c) (with (x, y, z) the solar-magnetospheric coordinates from Fig. 14.1f). Two neutral magnetic points where $B = 0$, called cusps, form on the dayside magnetopause near $\pm 75°$ geomagnetic latitude. Here, the solar wind has direct excess to the magnetosphere (e.g., Haerendel and Paschmann 1975). The solar wind, flowing around the sides of the magnetic barrier, stretches the geomagnetic field lines out into a long magnetic tail on the downwind night time side. This tail is several hundred earth radii long.

Actually, the outer magnetosphere is filled with a tennous low energy plasma, several orders of magnitude less dense than the solar wind plasma. The turbulent interaction between this plasma and the solar wind at the flanks of the tail induces a charge separation and creates an electric polarization field E (e.g., Cole 1974). This field just cancels the Lorentz field $v \times B$, so that the solar wind can stream undisturbed against the geomagnetic field in the boundary layer – the mantle in Fig. 10.3. This electric field (see Fig. 10.4) is of the order

$$E = - \Phi/\delta \simeq v B . \tag{10.3}$$

An electric potential of

$$\Phi_c = -2 \Phi \simeq 2 v B \delta \tag{10.4}$$

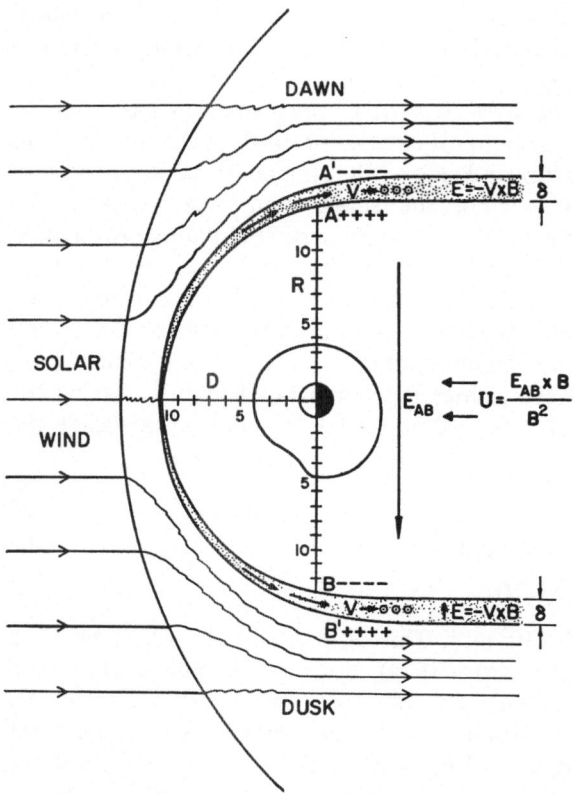

Fig. 10.4. Equatorial cross section of magnetosphere illustrating viscous-like interaction between solar wind and magnetosphere (Mendillo and Papagiannis 1971). The curve surrounding the earth with its bulge on the dusk side near five earth's radii distance is the plasmapause

will, then, be built up between points A and B in Fig. 10.4 with an electric field E_{AB}, directed from dawn to dusk within the equatorial plane. The thickness of the boundary layer δ is of the order of several hundred kilometers (e.g., Eastman et al. 1976), and the dawn-dusk potential generated by this viscous-like interaction is of the order of $\Phi_c \simeq 10\,\mathrm{kV}$.

Mapping of these electric potential fields along the geomagnetic lines of force into the magnetosphere is sketched in Fig. 10.5. The magnetic field lines crossing the inner border of the boundary layer and intersecting the earth near

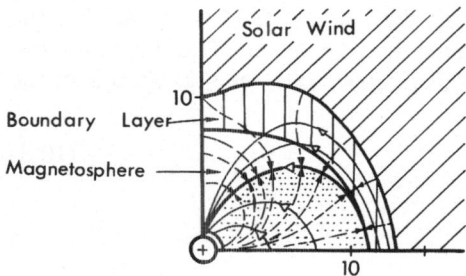

Fig. 10.5. Dawn-dusk cross section through magnetosphere as seen from the sun. *Solid arrows* indicate electric field lines. *Open arrows* indicate magnetic field lines. Plasma convection within the magnetosphere is directed from day to night in the polar regions and sunward within the inner magnetosphere

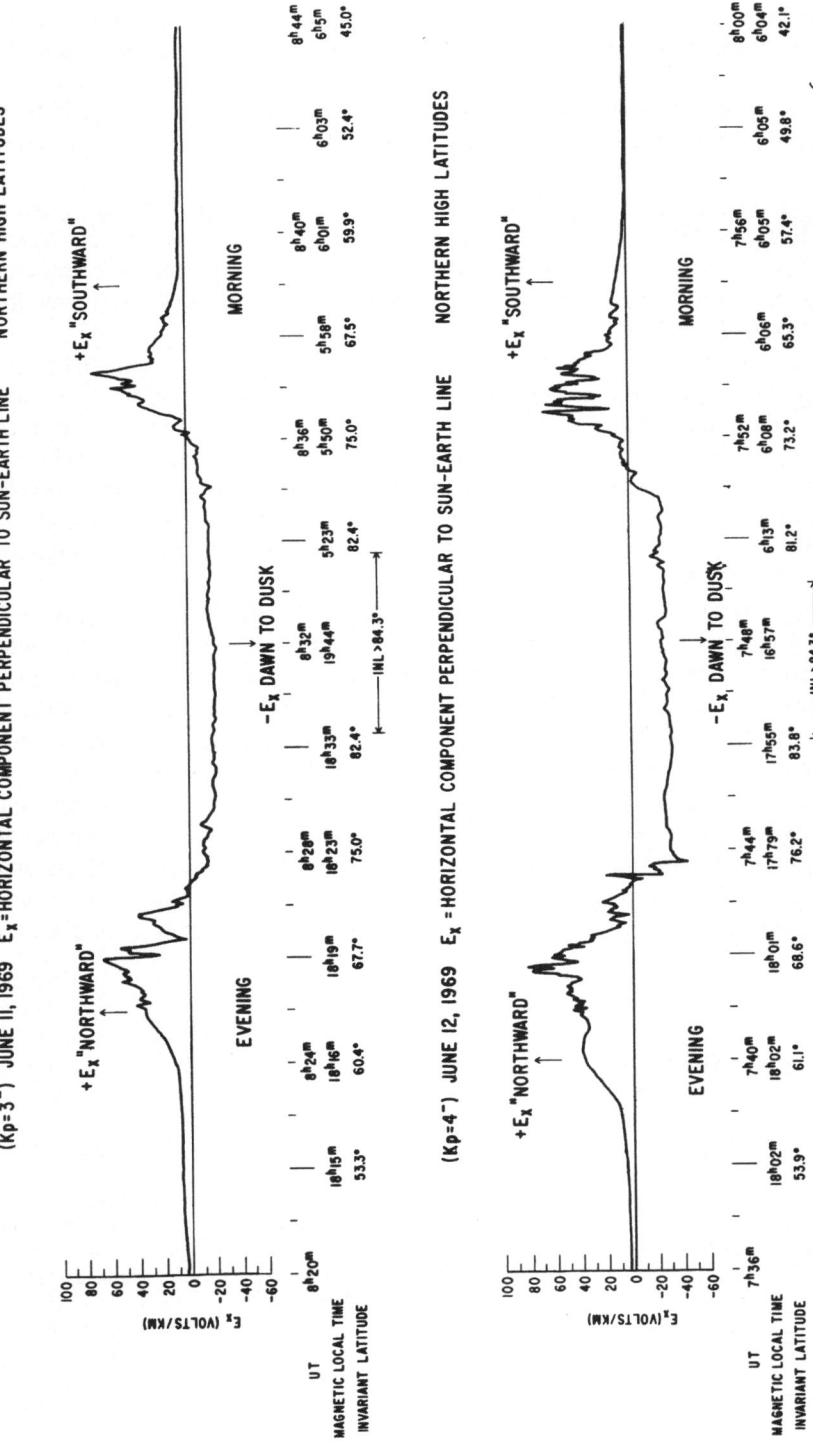

Fig. 10.6. Horizontal component of electric field perpendicular to the sun-earth line during two traverses of the OGO-6 satellite across the north magnetic pole. (Heppner 1972)

70° latitude separate two regions of opposite electric field direction. Electric fields of magnetospheric origin, which have this characteristic field reversal, have indeed been measured at ionospheric heights (Fig. 10.6). The magnetic field lines separating the two regions of oppositely directed electric field form an electric double layer in which field-aligned electric currents are expected to flow.

It follows from Fig. 10.5 that the $E \times B$ drift of the magnetospheric plasma must be toward the night side at high latitudes and toward the day side at low latitudes. A huge cell of plasma convection is therefore expected in the magnetosphere. The associated electric field is called the magnetospheric convection field (Axford 1969).

The magnetic field lines in the far tail are stretched nearly parallel to the ecliptic plane into space. The dawn–dusk-directed electric convection field forces the magnetospheric plasma in the far tail to drift toward the magnetically neutral sheet from both hemispheres (Fig. 10.7). In the neutral sheet where $B_x = 0$, the electrons move toward dawn while the protons move toward dusk, thus establishing a cross-tail current in the sheet directed toward dusk. The current is closed in the tail boundary layer above and below the magnetic lobes (Fig. 10.3).

Since magnetic field lines can be pictured as frozen into the highly conducting plasma, the plasma can be considered as transporting the magnetic field lines of opposite polarity toward the neutral sheet where they become continuously annihilated. This process is called field line reconnection (Sonnerup 1979). It is this mutual interaction between plasma and electric and magnetic fields which maintains the magnetic field topology in the tail.

Geomagnetic field lines having both foot prints on the earth (closed field lines) exist in the inner magnetosphere. Highly energetic ionized particles can be captured in certain zones of these closed field lines. The particles spiral along field lines from one mirror point to the other, and they drift in azimuthal direction, the protons to the west, the electrons to the east, thus estab-

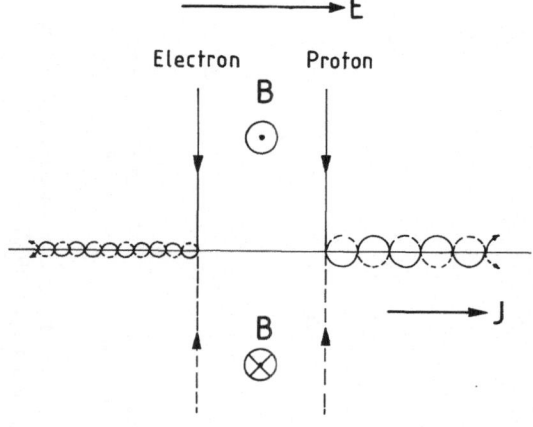

Fig. 10.7. Schematic illustration of magnetic reconnection process within the magnetospheric tail region. Movement of electrons and protons toward the neutral sheet of the tail under the influence of the dawn-dusk electric field E leads to an electric sheet current directed from dawn to dusk

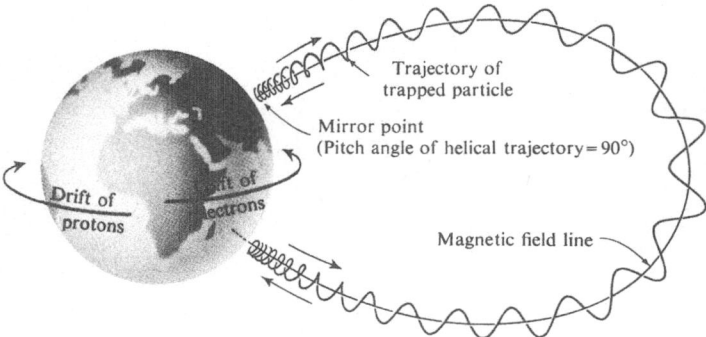

Fig. 10.8. Spiral motion and drift of protons and electrons within the geomagnetic field

lishing a zonal ring current (Fig. 10.8). Several belts of highly energetic particles (the van Allen belts) have been detected (Schulz and Lanzerotti 1974). However, the ions in the $10-200$ keV range which occupy a region at a distance between about 3 and 10 earth radii from the center of the earth contribute mainly to this ring current (Fig. 10.3).

10.3 Open Magnetosphere

In the previous section, we neglected the interplanetary magnetic field (IMF) carried by the solar wind. The field lines of the IMF merge with the geomagnetic field lines within the polar regions when the solar wind sweeps across the magnetosphere. A magnetosphere where the polar geomagnetic field lines are connected with the IMF lines is called an open magnetosphere. Figure 10.9 schematically shows a dawn − dusk cross section of the magnetosphere during away polarity (left) and toward polarity (right) of the IMF. It is well known that electric fields depend on the coordinate system in which they are observed (e.g., Alfven and Fälthammar 1963). If the solar wind is free of electric fields for an observer moving with the solar wind, an observer fixed in the frame of reference of the magnetosphere measures an electric field given by [Eq. (14.20)]

$$E_s = -v_s \times B_s,\qquad(10.5)$$

where v_s is the velocity of the solar wind and B_s is the IMF. This electric field is accompanied by electric polarization charges. Since the electric conductivity orthogonal to the field lines is nearly zero, no discharging current can flow in the undisturbed solar wind. When the solar wind reaches the earth's magnetosphere, field-line merging as indicated in Fig. 10.9 occurs, and the electric field E_s maps into the polar regions of the magnetosphere. Different IMF lines are now short-circuited via the ionospheric dynamo region, and a discharging

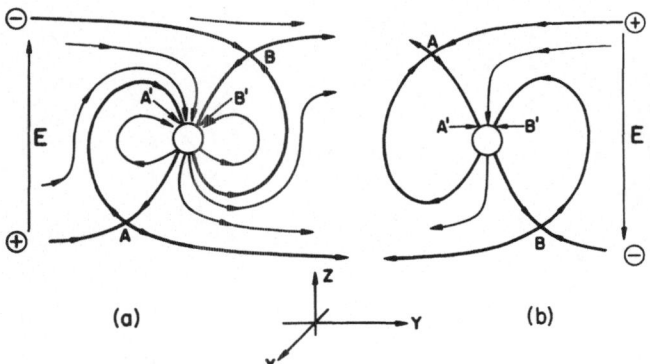

Fig. 10.9a, b. Schematic view of a dipole immersed in the interplanetary magnetic field with the solar wind moving into the plane of the drawing. **a** External magnetic field parallel to the y-axis (toward dusk) corresponding to away polarity of the IMF. **b** External field antiparallel to y-axis corresponding to toward polarity of IMF (Stern 1973)

current can flow, for instance, from point A to point B via A' and B' in Fig. 10.9.

The solar wind moves the magnetospheric plasma in the polar regions downstream. It also sweeps the rotating geomagnetic field lines to the night side where they again separate from the IMF and reconnect with each other. This process is accompanied by magnetospheric plasma convection downstream at high latitudes and upstream in the regions of closed field lines. The associated electric convection field is similar in shape to the convection field generated by the viscous-like interaction at the flanks of the tail, as discussed in Section 10.2. The magnetic field line where the electric field reversal occurs is the last closed field line. A quantitative approach to the convection field is given in Section 11.2.

The direction of the IMF has some influence on the electric convection field. If the IMF contains a south component (in the solar-magnetospheric co-ordinate system; see Fig. 14.1f), the electric convection field increases. If the vertical component of the IMF is northward-directed, the electric convection field tends to reach a ground state level. A south component of the IMF makes the magnetosphere more accessible to solar wind energy coupling. A north component tends to prevent such coupling. Moreover, as seen in Fig. 10.9, the azimuthal component of the IMF adds an antisymmetric field (with respect to the equator) to the convection field. We call this additional field the polar cap field. It will be discussed in Section 11.3.

We simulate the generation of the convection field within an open magnetosphere by a simple model of plasma flow through a channel in analogy to a magnetohydrodynamic (MHD) generator (e.g., Cap 1976) (Fig. 10.10). The channel (simulating the northern polar cap regions) has the dimensions (h, b, d). A steady-state flow of plasma [simulating the solar wind within the magnetosheath where its velocity has decreased by more than a factor of two

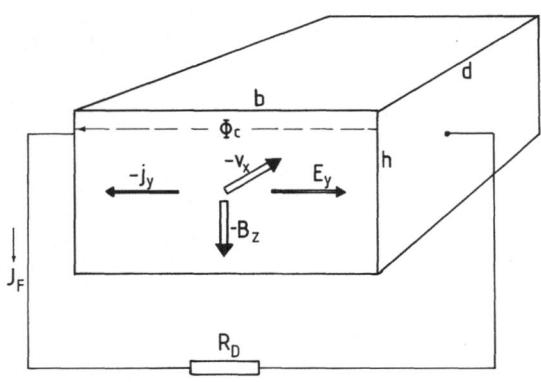

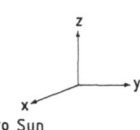

Fig. 10.10. Schematic view of hydromagnetic generator simulating solar wind flow into the northern hemispheric polar cap of the magnetosphere

as compared with the undisturbed solar wind velocity (Rosenbauer et al. 1975)] streams through this channel in the negative x-direction with a velocity $-v_x$. An external magnetic field, $-B_z$ (simulating the merged IMF and geomagnetic field), is southward-directed. The side walls of the channel are electrically connected by a load resistance R_D (simulating the electric contact between the IMF lines via the ionospheric dynamo region). An electric field in the y-direction E_y, and an electric current in the negative y-direction $-j_y$ are then induced, and a discharging current J_F flows through the ionospheric dynamo region. Ohm's law then demands

$$-j_y = \sigma_p(E_y - v_x B_z) . \qquad (10.6)$$

Furthermore, the current via the dynamo region of the ionosphere is related to the dawn−dusk potential Φ_c as

$$- \Phi_c = bE_y = hdR_D j_y = R_D J_F , \qquad (10.7)$$

from which it follows

$$j_y = \sigma_p v_x B_z/(1 + \sigma_p \delta)$$
$$E_y = \sigma_p \delta v_x B_z/(1 + \sigma_p \delta) , \qquad (10.8)$$

where $\delta = R_D hd/b$, R_D is from Eq. (2.21), and σ_p is the Pedersen conductivity in the channel orthogonal to B.

With the numbers $B_z = 20\,\text{nT}$, $v_x = 150\,\text{km/s}$, $h = b = d = 30a = 2 \times 10^5\,\text{km}$, $R_D = 30\,\text{m}\Omega$, $\sigma_p = 1.9 \times 10^{-8}\,\text{S/m}$, one arrives at $\Phi_c = 60\,\text{kV}$, $E_y = 0.3\,\text{mV/m}$, $J_F = 2\,\text{MA}$, $j_y = 50\,\text{pA/m}^2$. These numbers simulate moderately disturbed conditions. The electric power of the two MHD generators in both hemispheres is

$$P_{\text{mhd}} = 2J_F \Phi_c \approx 0.24\,\text{TW} . \qquad (10.9)$$

This may be compared with the total kinetic power of the plasma within both channels of

$$P_{\text{kin}} \approx bhd\varrho_s v_x^2/\tau \approx 2.8\,\text{TW} , \qquad (10.10)$$

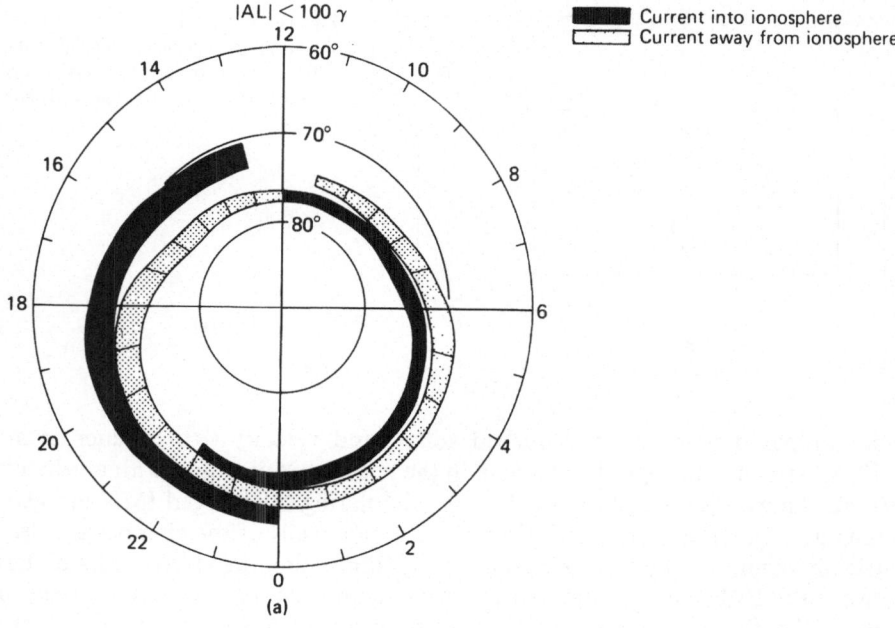

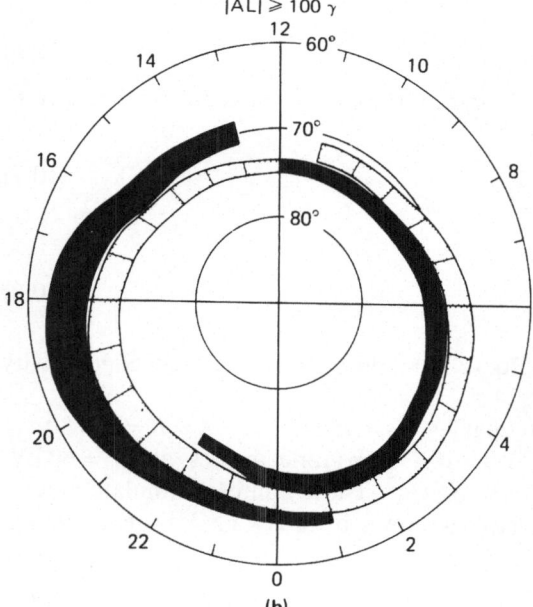

Fig. 10.11a, b. Distribution and flow direction of large scale field-aligned (Birkeland) currents during weakly disturbed conditions (*above*), and during active periods (*below*) on the northern hemisphere. Region I is at the polar side, region II is at the equatorward side (Iijima and Potemra 1978)

where $\tau \simeq d/v_x \simeq 1300$ s is the residence time of the plasma within the channels, and $\varrho_s \simeq 2 \times 10^{-20}$ kg/m^3 is the solar wind density in the magnetosheath. Thus, nearly 10% of the solar wind energy is converted into electric energy. Since the cross-tail currents and the boundary currents have the same order of magnitude as the field-aligned currents, total energy dissipation of the solar wind into the whole magnetosphere is larger.

Akasofu (1980) developed the following semi-empirical formula for the energy dissipation into the magnetosphere (in W):

$$P_{mag} \simeq [B_s^2/(2\mu)]\, v_s F_s \sin^4(\alpha/2) \tag{10.11}$$

where v_s (in m/s) is the solar wind speed, B_s (in T) is the strength of the IMF, $F_s \simeq (20\,a)^2\, \pi \simeq 16\,\pi\, 10^{15}$ m^2 is a typical cross-section of the magnetosphere, and α is the angle between the IMF and the geomagnetic dipole field (see Fig. 10.15). It is $\alpha = 0$ for purely northward directed IMF, and $\alpha = \pi$ for purely southward directed IMF. P_{mag} may vary from 0.1 TW during quiet conditions to 10 TW during large disturbances.

Dawn – dusk potential differences increasing from about 20 kV during quiet conditions to more than 100 kV during strongly disturbed conditions (Reiff et al. 1981), and field-aligned electric currents (Birkeland currents) of the order of a few megaamperes flowing into the ionosphere on the morning side and flowing out on the evening side at auroral latitudes (region I in Fig. 10.11) have indeed been observed.

The region I currents in Fig. 10.11 delineate the boundary of the polar cap from which the open field lines originate. They roughly determine the area of the auroral oval where luminous events – the auroral lights – can be observed during nearly every clear night. Auroral light is deactivation luminosity from neutrals and ions in excited states. The neutrals are excited primarily by energetic electrons in the 1 to 10 keV range precipitating from the magnetospheric tail regions on the night side and from the cusp region on the day side

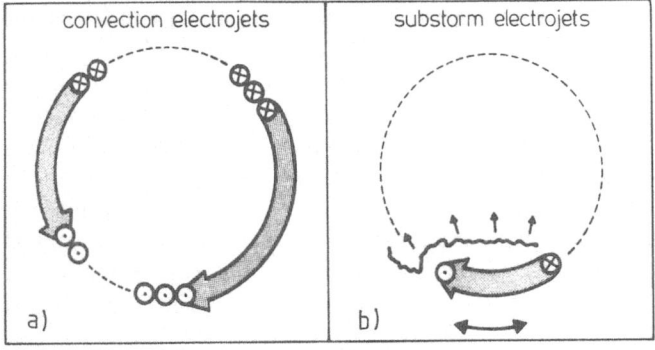

Fig. 10.12a, b. Illustrative view of the Hall components of the polar electrojets and field-aligned current closure during quasi-stationary conditions (a), and transient substorm electrojet (b) (Baumjohann 1983)

into the auroral ionosphere. Precipitating electrons and protons of higher energy can ionize the neutrals so that the electric conductivity in the auroral ovals is increased, and part of the field-aligned electric currents are closed directly via this region of enhanced electric conductivity. The two strong bands of ionospheric electric currents in the auroral dynamo region of both hemispheres are called the auroral or polar electrojets (Fig. 10.12a). Auroral lights as well as the polar electrojets increase in strength with increasing geomagnetic activity (e.g., Aksofu 1977).

An increase in the southward component of the IMF also increases the azimuthal electric field component E_y via Eq. (10.8) and thus increases the dawn-dusk potential. This is in agreement with observations. Presently, it is believed that the mechanism of viscous-like interaction, as discussed in Section 10.2, as well as the MHD generator mechanism, contribute jointly to the magnetospheric electric convection field. During quiet conditions, both mechanisms may supply a comparable amount of energy into the magnetosphere. The MHD generator probably becomes more important during disturbed conditions (Hones 1979a).

10.4 Geomagnetic Activity

The superposition of magnetic fields originating from all fluctuating magnetospheric electric currents (ring current, field-aligned currents, ionospheric currents, cross-tail currents, boundary currents) measured on the ground determines the degree of geomagnetic activity. Geomagnetic activity is characterized by various indices measuring the degree of disturbance (e.g., Rostoker 1972; Mayaud 1980). One index − the a index − is a linear measure of the largest amplitude in the horizontal component of the geomagnetic field departure from the regular Sq variation during a 3-h interval. The ap index is the worldwide average of the a index from several selected stations. Daily indices can be computed from the 3-h indices. For example, the Ap index is the average ap index over one UT-day. An improved ap index derived from selected stations separated into stations in the northern and in the southern hemisphere is called the am index. Expressing the geomagnetic disturbances on a logarithmic scale from 0 (very quiet conditions) to 9 (most disturbed conditions) gives a representation of geomagnetic activity on a 3-h basis − the K index and a corresponding world wide index Kp. Another widely used index is the AE index. This index is derived from stations located near the auroral oval and reflects mainly the strength of the auroral electrojets (e.g., Kamide and Akasofu 1983). Disturbances at low latitudes are characterized by the Dst index.

The origin of the strongly fluctuating geomagnetic field on the ground (see Fig. 8.1b) is the variability of the solar wind impinging on the magnetosphere, but also the systematically varying geometric configuration of the sun−earth

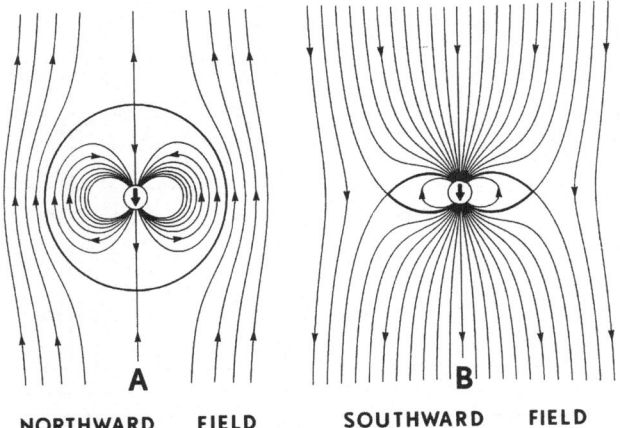

NORTHWARD   FIELD   SOUTHWARD   FIELD

Fig. 10.13 A, B. Magnetic field lines of the sum of a dipole field and a homogeneous field directed antiparallel to the dipole axis (**A**), and parallel to the dipole axis (**B**) (Hill and Rassbach 1975)

system. The main solar wind parameter is the azimuthal electric field [Eq. (10.8)]. Electric currents flow within the magnetosphere in order to maintain the magnetospheric tail. Any change of the electric currents reacting to solar wind fluctuations forces the tail to reach a new equilibrium.

A very simple magnetic field configuration simulating the magnetosphere and the IMF is the magnetic field topology of Fig. 10.13: a dipole field imbedded in a homogeneous magnetic field. In Fig. 10.13a, the axis of the dipole is directed opposite to the homogeneous field. In Fig. 10.13b, it has the same direction as the homogeneous field. The left panel resembles a closed magnetosphere, the right panel an open magnetosphere.

If the dipole were to move freely, it would turn into the configuration of Fig. 10.13b like a compass needle in the geomagnetic field. In order to return the dipole to the configuration of the left panel (or turn a compass needle by 180°), mechanical work of $2MB_z$ has to be applied, where M is the geomagnetic dipole moment ($M = 8 \times 10^{22}$ Am^2) and B_z is the strength of the homogeneous field. If the IMF turns southward again, mechanical energy is released which is available to heat the tail plasma up to

$$W_{\text{tail}} \simeq 2MB_z \simeq 5 \times 10^{14}\,\text{J} = 500\,\text{TJ} \tag{10.12}$$

(with $B_z \simeq 3$ nT the southward component of the IMF).

The prototype of an isolated disturbance is a magnetospheric substorm. Two substorm models are presently proposed (Fig. 10.14). In the reconnection model (upper panel), one assumes that solar wind energy enters the magnetosphere after the IMF has turned southward. The electric currents in the magnetosphere are enhanced and stretch the field lines in the tail downstream. This force is counteracted by magnetic tension which tries to shorten the magnetic field lines. This is an instable field configuration leading to the development of a magnetic neutral line (X-line). Unloading of the stored

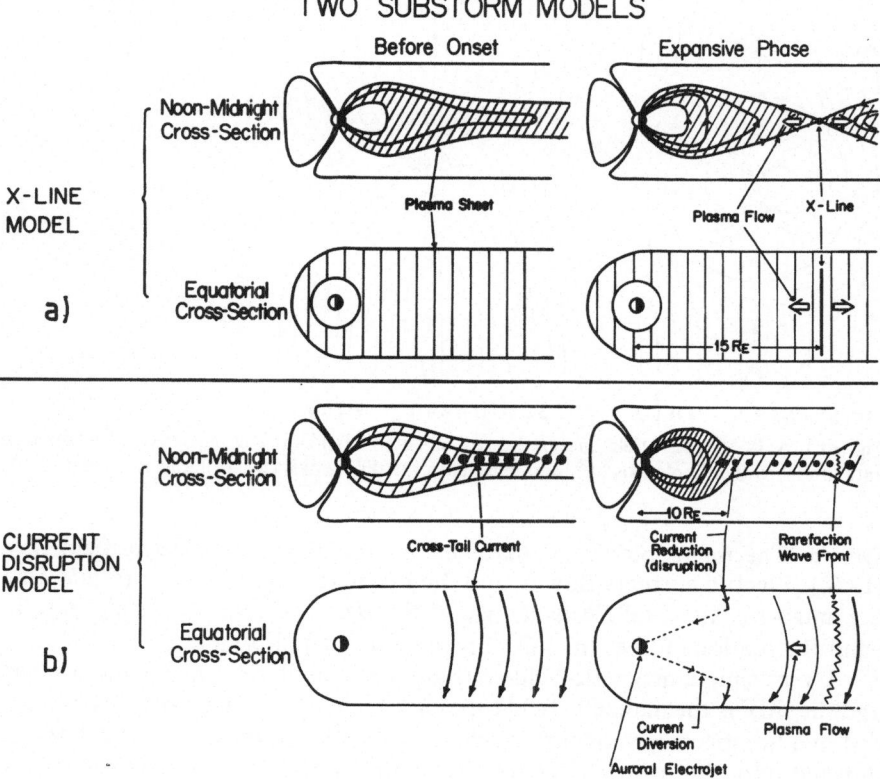

Fig. 10.14 a, b. Two views of the cause of magnetospheric substorms. *Upper panel* the reconnection model is shown in the noon-midnight cross section and the equatorial cross section. *Lower panel* shows the current disruption model, illustrated in the same way (Akasofu 1980). The locations of the *X*-line (at 15 R_E) and of the current disruption (at 10 R_E) in the tail should be considered as typical values which may vary widely in each individual case

energy occurs in a cataclystic way. Hereby, a blob of magnetized plasma may become detached from the magnetotail plasma and may be carried away downstream by the solar wind while the reconnected inner field lines contract toward the earth (Hones 1979b; Schindler 1980). During the contraction of the near tail field, electrons and ions in these regions are heated to energies above 1 keV. Some of these energetic particles stream along the magnetic field lines and then precipitate into the auroral ovals of the ionosphere, causing the auroral lights and the zones of enhanced electric conductivity. The heating and acceleration processes are not yet well understood. It is believed that field-aligned electric potentials of the order of a few kilovolts may be involved in the acceleration process (e.g., Mozer 1981).

Other energetic particles enhance the ring current population in the magnetosphere, shifting its center toward the earth (Williams 1983). The ring current is stronger on the night side than on the day side, so that part of it must be closed via field-aligned currents and via the ionosphere. The region II field-aligned currents in Fig. 10.11 may be these discharging currents from the inner magnetosphere. The partially closed ring current is accompanied by accumulation of electric polarization charges on the morning and evening side in the inner magnetosphere, which weaken the electric convection field in this area (e.g., Stern 1983).

The essential idea of the second substorm model (lower panel in Fig. 10.14) is that the cross-tail current diverts after the IMF has turned southward. A shunt current consisting of field-aligned currents and the substorm electrojet on the night side (Fig. 10.12b) develops, driven by the enhanced electric convection field, while the earthward-streaming tail plasma fills the ring current. When the IMF turns northward again, the magnetic energy stored in the tail [Eq. (10.12)] is released in an abrupt way (expansion phase of the substorm). We will discuss this model in more detail in Section 11.7.

A single event of the kind just described typically lasts up to a few hours. The additional westward-directed current system in Fig. 10.12b in the night time auroral oval leads to geomagnetic disturbances of the order of several 100 nT on the ground below the substorm electrojet. This geomagnetic disturbance is called a geomagnetic substorm or a DP1 event (Kamide and Baumjohann 1984). A series of intense substorms, lasting for 1 or more days, is called a magnetospheric storm (Akasofu 1977).

10.5 Seasonal and Universal Time Variations of Geomagnetic Activity

Although magnetospheric storms and substorms appear to be rather complex phenomena which are by no means understood in every detail, their magnetic effects on the ground seem to have a consistent one-to-one relationship with the solar wind parameters, at least on a statistical basis. This indicates that the magnetospheric processes may be driven by the solar wind and are not instable unloading phenomena (Akasofu 1981).

Solar wind parameters influencing the geomagnetic activity are the momentum flux $m_s n_s v_s^2$, and the north–south component of the electric Lorentz field $v_s B_s \cos \alpha$, where $B_z = B_s \cos \alpha$ is the north–south component of the IMF in the solar–magnetospheric coordinate system and α is defined in Fig. 10.15. Since the earth's axis is inclined by 23.5° to the ecliptic plane, and the geomagnetic axis is inclined by 11.2° to the earth's rotation axis, the angle A between the geomagnetic dipole axis and the ecliptic plane can vary from 55° to 125° depending on season and universal time. Furthermore, the heliographic equator is inclined by 7.25° to the ecliptic plane, so that B_z varies systematically as a function of season, universal time, and the solar 22-year cycle.

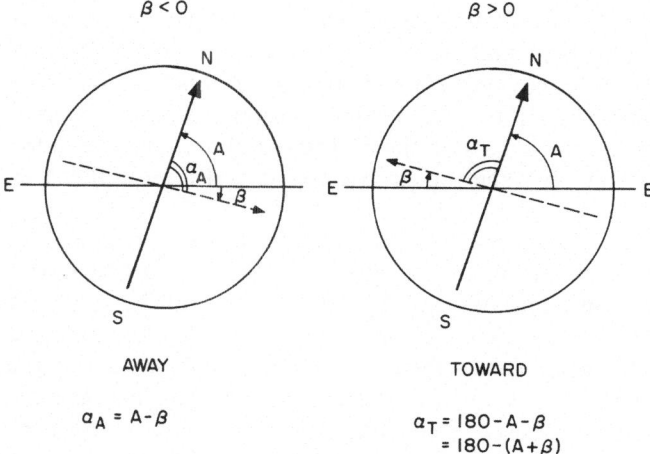

$$\alpha_A = A - \beta$$

$$\alpha_T = 180 - A - \beta$$
$$= 180 - (A + \beta)$$

Fig. 10.15. Field line geometry at the nose of the magnetosphere as seen from the sun for toward polarity of the IMF (*right*), and for away polarity of the IMF (*left*). The direction of the IMF is indicated by the *dashed arrows*. The projection of the geomagnetic field is indicated by *solid arrows*. The ecliptic plane is along *EE*. The angle *A* is the angle between the ecliptic and the geomagnetic field projection. The angle α is defined as shown separately (Svalgaard 1977)

Svalgaard (1977) developed the following empirical relation between the planetary *am* index and the solar wind parameters:

$$am \simeq 3 \times 10^{-4} v_s B_s (n_s v_s^2)^{1/3} v_s^{1/4} \exp(-0.74 \cos \alpha)/(1 + 3 \cos^2 \psi)^{2/3} \quad (10.13)$$

(v_s in m/s; B_s in T; n_s in m^{-3}). The influences of the momentum flux proportional to $n_s v_s^2$, the Lorentz field $v_s B_s$, and the direction of the IMF vector α, are separated in Eq. (10.13). The factor proportional to $v_s^{1/4}$, accounts for solar wind fluctuations. The tilt of the dipole field with respect to the ecliptic plane is included in Eq. (10.13) by the factor in the denominator. The angle between solar wind direction and the dipole axis is ψ, and

$$B_0 = B_{eq}(1 + 3 \cos^2 \psi)^{1/2} \quad (10.14)$$

is the dipole field strength at the subsolar point with B_{eq} the field strength at the geomagnetic equator.

The dipole tilt modulation of *am* in Eq. (10.13) is a purely empirical result and lacks a clear physical explanation. Boller and Stolov (1970) have suggested that it is due to instabilities arising from the relative motion of the two MHD fluids at their boundary (Kelvin-Helmholtz instability). It is also possible that the coupling between solar wind and magnetosphere depends on the basic geometry of the magnetic field around a dipole.

Figure 10.16a shows a plot of

$$am = am_0(1 + 3 \cos^2 \psi)^{-2/3} \quad (10.15)$$

versus universal time and season, with $am_0 = 24$. Figure 10.16b gives the contour plot of the observed variation of the *am* index based on the average of 16

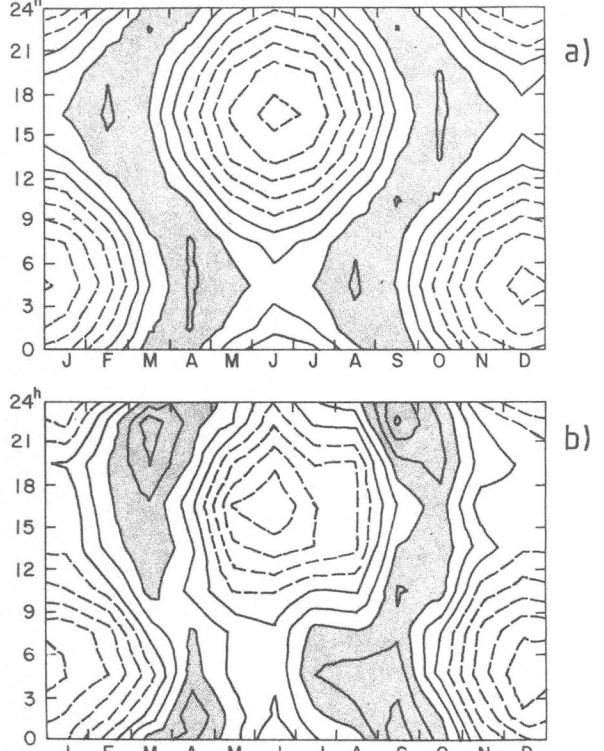

Fig. 10.16a, b. Contour plot of *am* computed from Eq. (10.15) (*upper panel*); and contour plot of observed variation of the *am* index with time of year and universal time, based on data from 1959 – 1974 (*lower panel*). The first full contour line corresponds to *am* = 21, and the lines are drawn one *am*-unit apart. Contour lines in regions of lower than average are shown as *dashed lines* (Svalgaard 1977)

years of data (1959 – 1974) so that the variability of the solar wind parameters in Eq. (10.13) is smoothed out. The agreement is good, indicating that the dipole tilt modulation is responsible for a semiannual variation of *am* with maximum amplitudes during equinox where B_0 in Eq. (10.14), averaged over the day, has its minimum value. Furthermore, *am* exhibits a diurnal UT variation with extreme amplitudes at solstices during the morning and the evening hours. These UT variations are in antiphase during the June and the December solstices. The UT variation displays a small amplitude semidiurnal wave during the equinox.

Averaged over all seasons, the resulting UT variation is small and hardly detectable. However, if the days are divided into days of opposite IMF polarity, seasonally averaged UT variations are clearly discernable with opposite phases for toward (*T*) and away (*A*) polarity of the IMF. This is seen in Fig. 10.17 where the difference $am(A) - am(T)$ and the sum $[am(A) + am(T)]/2$ are plotted versus UT for December solstice, equinox, and June solstice. We notice that the polarity-dependent UT variation is essentially independent of season, while the polarity-independent variation displays the effect shown in Fig. 10.16.

The variation of geomagnetic activity with the 11-year cycle is shown for the last four sunspot cycles in Fig. 10.18. The daily index *Ap*, typically, shows

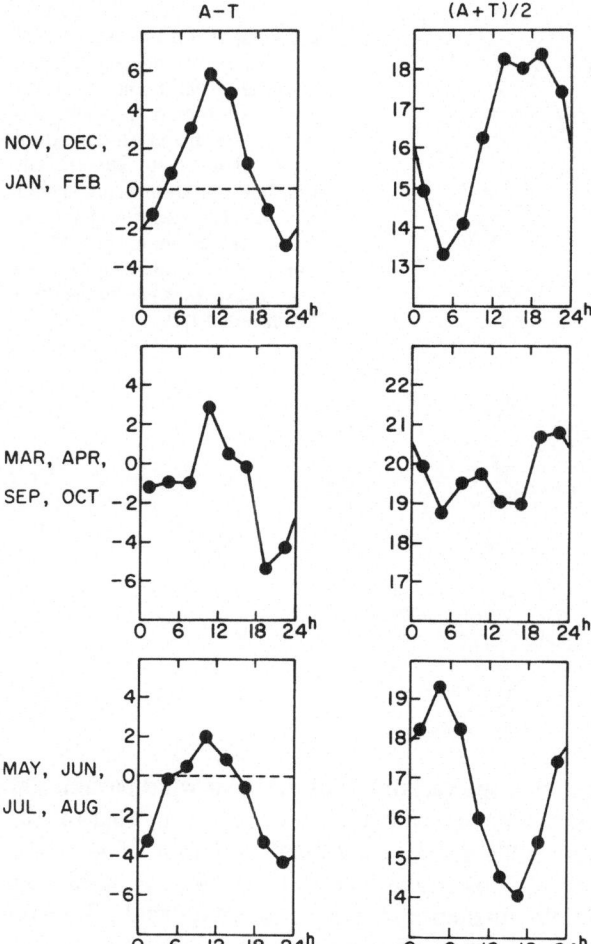

Fig. 10.17. Universal time dependence of the difference between the *am* index on away-days (*A*) and on toward-days (*T*) (*left*), and average (*A* + *T*)/2 (*right*) for different seasons (Svalgaard 1975)

a relative maximum shortly before or during sunspot maximum and an absolute maximum 4 to 5 years later. The first maximum of *Ap* results, essentially, from the relatively small number of large geomagnetic storms associated with solar flare-induced plasma emissions into the solar wind. The absolute maximum in *Ap*, which occurs after the sunspot maximum, is related to high speed solar wind streams from coronal holes. The life times of coronal holes have their maximum during the declining phase of solar activity. Therefore, recurrent moderate geomagnetic storms are initiated, which sum up to raise the *Ap* index to its absolute maximum (Siebert 1977; Feynman 1983).

A small 22-year cycle effect of geomagnetic activity was detected by Chernosky (1966) and was interpreted by Russell (1974) as follows: The earth is at its highest northern heliographic latitude (7.25°) during September and at its highest southern heliographic latitudes (−7.25°) during March. Since the

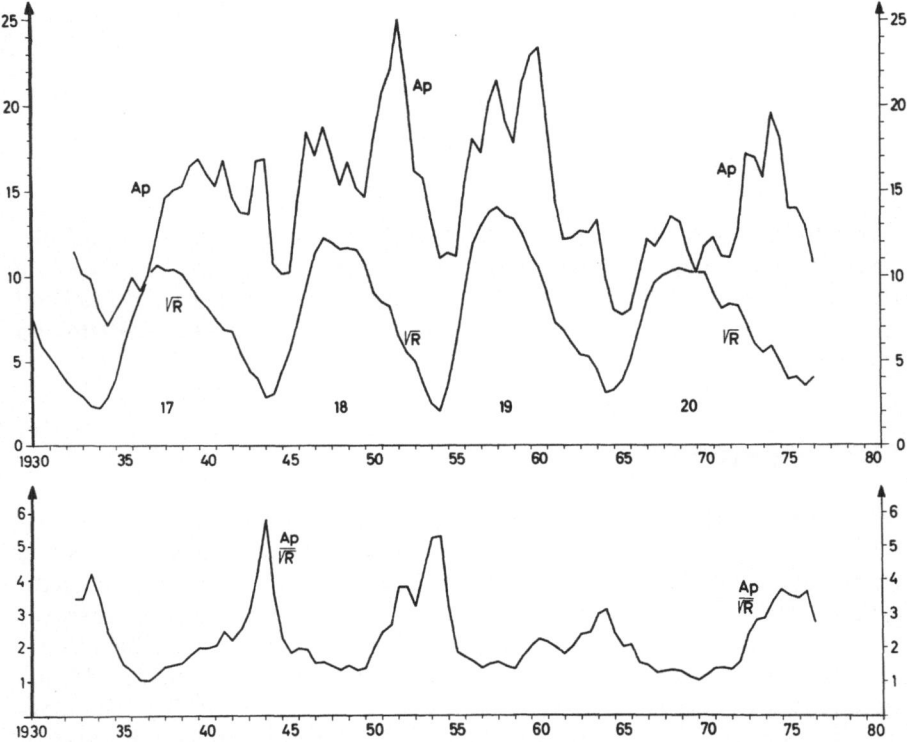

Fig. 10.18. Square root of sunspot number R, and geomagnetic activity number Ap during the four sunspot cycles 17–20 (*upper panel*); and quotient $Ap/\sqrt{R}$ (*lower panel*). The quotient $Ap/\sqrt{R}$ can be interpreted as a measure of the life time of the coronal M-regions (Siebert 1977)

IMF in the northern heliographic hemisphere has, on the average, one polarity, and in the southern hemisphere the opposite polarity (see Fig. 10.1), the earth tends to spend more time south of the heliographic current sheet during March equinox and north of this sheet during September equinox (Wilcox and Scherrer 1972).

An average negative B_z accompanied by higher geomagnetic activity, is therefore expected during a cycle where the solar dipole axis is northward-directed. An average positive B_z, accompanied by lower geomagnetic activity, is expected during a cycle where the solar dipole axis is southward-directed. The north pole was in the northern hemisphere from 1968–1979 so that geomagnetic activity was presumably higher than the previous 11 years (1957–1968) where the north pole was in the south.

11 Global Magnetospheric Electric Fields and Currents

In the foregoing chapter, we made simple estimates of electric fields and currents within the magnetosphere. We quantify in this chapter the global electric field and current configuration within the inner magnetosphere ($r < 15a$) and its magnetic manifestation on the ground [see also Gurevich et al. (1976) and Pudovkin (1974)].

11.1 Convection Field

In order to calculate global magnetospheric electric fields, we will rely on extremely simplified models. In particular, we restrict ourselves to quasi-static electric potential fields, and we make no pretense of solving for the magnetospheric plasma convection and its electromagnetic field in a self-consistent manner. We eliminate the MHD generator by introducing plausible electric field configurations and calculate the electric currents from the E-field, assuming that the electric conductivity is known (Vasyliunas 1970). The magnetic fields of the currents are expected to be small compared with the internal geomagnetic field (approximated in our approach by a dipole field). This is certainly true within the inner magnetosphere.

The lower boundary of the magnetosphere is the dynamo region, approximated by a thin shell at height $r_D \simeq a$ [Note that $(r_D - a)/a \simeq 0.016$] with a constant height integrated electric conductivity tensor Σ [Eq. (2.20)] in which horizontal currents flow:

$$J = \Sigma \cdot (- \nabla \Phi + U \times B_0) . \tag{11.1}$$

Φ is the electric potential of the convection field. U is a horizontal neutral wind generated by Joule heating due to the electric current. The dynamo action of this wind modifies the electric currents and fields. This modification is, however, a second-order effect and will not be considered here (e.g., Blanc and Richmond 1980).

Since no sources or sinks of the electric current exist in the inner magnetosphere and in the ionosphere, the condition $\nabla \cdot j = 0$ implies field-aligned currents j_F flowing from the top of the dynamo region into the magnetosphere:

$$j_F = - \nabla \cdot J / \sin I , \tag{11.2}$$

with I the dip angle of the geomagnetic field.

Clearly, the field-aligned currents, the dynamo currents, and the electric convection potential are all driven by the magnetospheric MHD generator and adjust to one other. In the semi-empirical approach which we adopt, one may select observed field-aligned currents such as in Fig. 10.11 in order to determine the dynamo currents and the convection potential (Yasuhara and Akasofu 1977; Nisbet et al. 1978; Kamide and Matsushita 1979), or one may start from the observed electric field such as in Fig. 10.6 to determine electric potential and currents.

We follow the second approach and evaluate first the possible configurations of the electric convection potential. For this reason, we start from a magnetic field configuration of the magnetosphere that is composed of a dipole field on which a homogeneous field as in Fig. 10.13b is superposed. The magnetic components of this field in spherical coordinates are (Hill and Rassbach 1975; Schulz 1976):

$$B_r = [-2B_{00}(a/r)^3 - B_i]\cos\theta$$
$$B_\theta = [-B_{00}(a/r)^3 + B_i]\sin\theta \tag{11.3}$$
$$B_\lambda = 0.$$

B_{00} is the dipole field at the equator on the ground [Eq. (14.4)], and B_i is the magnitude of the homogeneous field. The equation of the field lines can be derived from the differential equation $dr/(r\,d\theta) = B_r/B_\theta$ as

$$L = r/(a\sin^2\theta)/[1 + 0.5(r/b)^3] = \text{const.} \tag{11.4}$$

L is a shell parameter. In the case of no external field ($B_i = 0$), the shell parameter becomes that of a dipole field [Eq. (14.9)]. Furthermore, the parameter in Eq. (11.4)

$$b = a[B_{00}/B_i]^{1/3} \tag{11.5}$$

is the equatorial distance of the last closed field line. The shell parameter of the last closed field line is

$$L_b = 2b/(3a). \tag{11.6}$$

L is related to the axial distance $\varrho_\infty = r\sin\theta$ at infinity ($r \to \infty$) as

$$L = 2b^3/(a\varrho_\infty^2) = 3L_b b^2/\varrho_\infty^2. \tag{11.7}$$

The axial distance of the last closed field line, according to Eqs. (11.6) and (11.7), is

$$\varrho_{b\infty} = (3)^{1/2}b \tag{11.8}$$

The shell parameter L_b, called the separatrix (Vasyliunas 1975), delineates three regions:

a) a region of closed field lines ($L < L_b$) simulating the inner magnetosphere. The foot prints of the lines L_b delineate the polar borders of the auroral zones;
b) a region of open field lines with one foot print on the earth ($L > L_b$;

$\varrho_\infty < \varrho_{b\infty}$), simulating the polar regions and the tail of the magnetosphere, and c) a region with field lines unconnected to the dipole ($L > L_b$; $\varrho_\infty > \varrho_{b\infty}$), simulating the interplanetary magnetic field.

If we assume an asymptotically uniform electric field E_y in the distant tail directed from dawn to dusk, mapping into the tail region (b) leads to a potential (expressed in solar-magnetospheric coordinates; Fig. 14.1f) given by Eqs. (11.7) and (9.34)

$$\Phi_c = E_y y = -E_y \varrho_\infty \sin\lambda = -(\Phi_{c0}/2)(L_b/L)^{1/2}\sin\lambda \quad (L > L_b) \tag{11.9}$$

with $\Phi_{c0} = (12)^{1/2} E_y b$.

The electric potential in the inner magnetosphere [region (a)] can be an arbitrary function of L [see Eq. (9.34)]. However, Φ_c must be continuous at L_b. We choose for convenience the form

$$\Phi_c = -(\Phi_{c0}/2)(L/L_b)^{q/2}\sin\lambda \quad (L < L_b). \tag{11.10}$$

From the observed plasmapause configuration (see Sect. 11.2), the number $q \simeq 4$ appears appropriate (e.g., Burke 1982).

For an observer on the earth, it is more convenient to transform from solar–magnetospheric coordinates to geomagnetic coordinates (Fig. 14.1e). This implies a replacement of the azimuth λ by the geomagnetic local time τ and $\sin\lambda$ by $-\sin\tau$. We neglect here the real configuration of the auroral ovals with their finite latitudinal extent of about 500 km and their centers displaced by about 4° toward the night side with respect to the geomagnetic axis.

The electric field components of Eqs. (11.9) and (11.10) are

$$E_r = -[q/(2r)][1-(r/b)^3]/[1+0.5(r/b)^3]\,\Phi_c$$

$$E_\theta = (q/r)\cot\theta\,\Phi_c \tag{11.11}$$

$$E_\lambda = -\cot\tau/(r\sin\theta)\,\Phi_c$$

with $q = -1$ for $L > L_b$ and $q = 4$ for $L < L_b$. Since the electric potential is symmetric with respect to the equator, we consider in the following only the northern hemisphere.

Figure 11.1 shows E_θ along the dawn meridian and E_λ along the noon meridian. Here, the numbers $L_b = 10$ and $\Phi_{c0} = 30\,\text{kV}$ have been applied, simulating slightly disturbed conditions ($Kp \simeq 2$). E_θ displays a field reversal at $\theta_b = 18.4°$ colatitude; and the field in the polar cap is nearly constant. This is qualitatively in accordance with the observations in Fig. 10.6. Figure 11.2 shows the equipotential lines of Φ_c within the ionosphere ($r = r_D \simeq a$). Figure 11.3 gives the equipotential lines of Φ within the equatorial plane of the inner magnetosphere ($\theta = 90°$; $L < L_b$). A total potential difference between dawn and dusk of $\Phi_{c0} = 30\,\text{kV}$ exists.

The field-aligned currents can be determined from Eqs. (11.1), (11.2) and (11.11) as

$$j_F = \Phi_c \Sigma_p (q-1)(q+1-q\sin^2\theta)/(a^2\sin^2\theta\sin I). \tag{11.12}$$

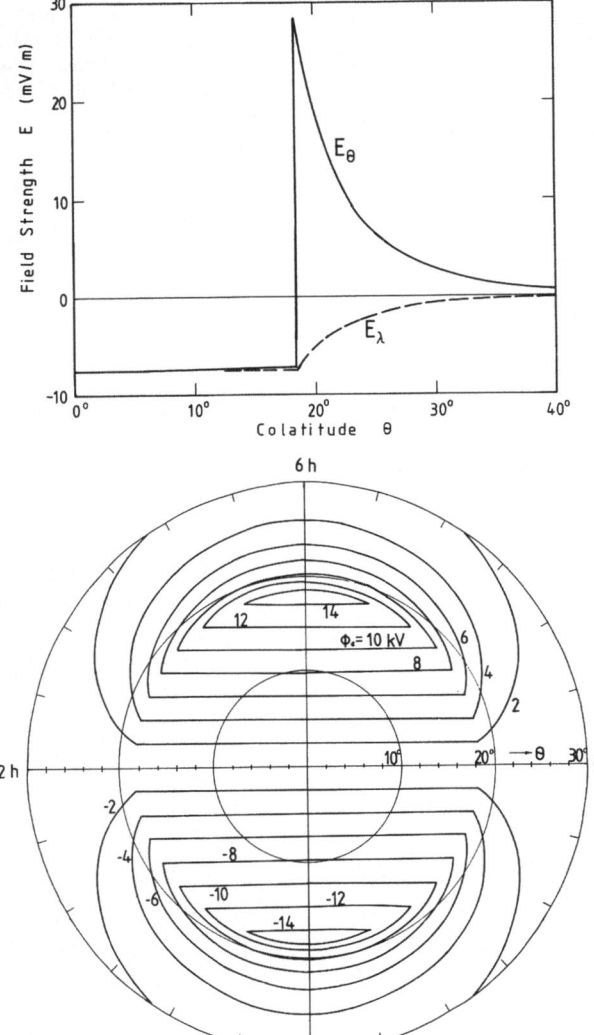

Fig. 11.1. Meridional electric field during dawn (*solid line*); and zonal electric field during noon (*dashed line*) of convection field during weakly disturbed conditions at ionospheric heights, as derived from Eq. (11.11)

Fig. 11.2. Equipotential lines of potential of electric convection field at ionospheric heights on the northern hemisphere during weakly disturbed conditions versus local time, as seen from the north downward [see Eqs. (11.9) and (11.10)]. Between two lines is a voltage difference of 2 kV

Since the Hall conductivity is constant in our approach, the field-aligned current consists only of the Pedersen component. The shell parameter L_b where the electric field reversal occurs is an electric double layer in which a field-aligned electric sheet current flows:

$$J_F = (J_{\theta p}\Big|_{\theta_b-0} - J_{\theta p}\Big|_{\theta_b+0})/\sin I = -(q+1)\Sigma_p\,\Phi_{c0}\cot\theta_b\sin\tau/(2a\sin I)\,,$$

$$(11.13)$$

with $\theta_b \simeq \arcsin(1/\sqrt{L_b})$ the colatitude of the polar border of the auroral oval in the northern hemisphere, and $J_{\theta p}$ the Pedersen component of J_θ.

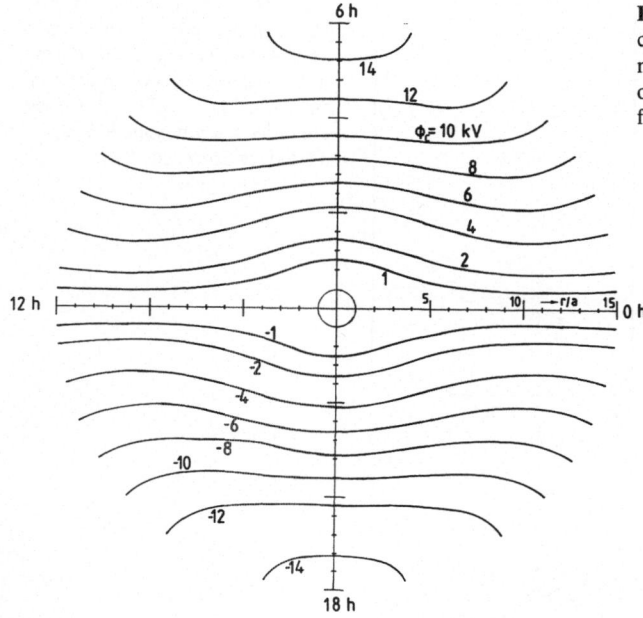

Fig. 11.3. Equipotential lines of convection field of Fig. 11.2, mapped into the equatorial plane of the magnetosphere, as seen from the north

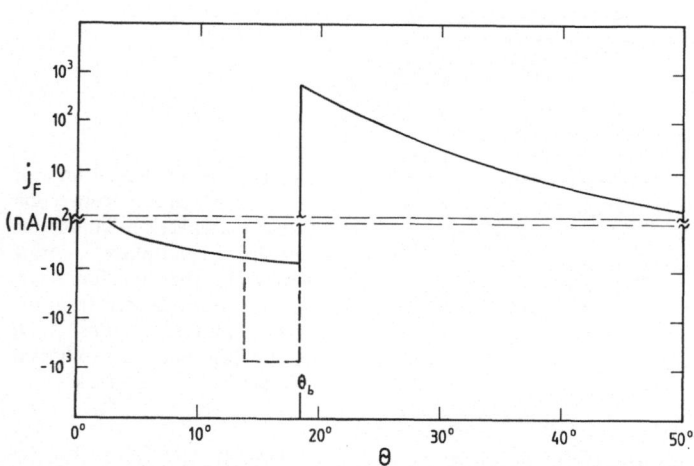

Fig. 11.4. Field-aligned current density during dawn, related to the electric field configuration of Figs. 11.1 to 11.3. The strength of the sheet current at the electric field reversal at θ_b is $J_F = -0.36$ A/m. *Dashed line* gives a fictive field-aligned current density of $J_F/\Delta x = 0.72$ μA/m^2 with $\Delta x = 500$ km a typical width of the auroral oval

Figure 11.4 shows the field-aligned currents along the dawn terminator derived from Eqs. (11.12) and (11.13) with inflow into the ionosphere in the polar region ($\theta < \theta_b$) and outflow for $\theta > \theta_b$. The flow direction reverses on the evening side. Here we used $\Sigma_p = 10$ S.

Comparison with Fig. 10.11 indicates that the field-aligned sheet current along L_b corresponds to the region I current in Fig. 10.11, while the currents at $L < L_b$ correspond to the region II current. The total current flowing into region I in the morning hemisphere is

$$J_{total} = a \sin \theta_b \int_0^{12h} J_F d\tau = -5 \Sigma_p \Phi_{c0} \cos \theta_b / \sin I \simeq -1.4 \, \text{MA} . \qquad (11.14)$$

A current of the same order flows out of region II.

The oversimplified field reversal of E_θ in Fig. 11.1 with its exaggerated peak and the infinitely thin sheet current in Fig. 11.4 can be brought into better agreement with the observations by adding a transition region within the auroral oval where E_θ changes continuously (Volland 1978).

The observations indicate that sometimes the electric field in the polar caps can better be represented by an exponent $q = -2$ instead of $q = -1$ (Heelis et

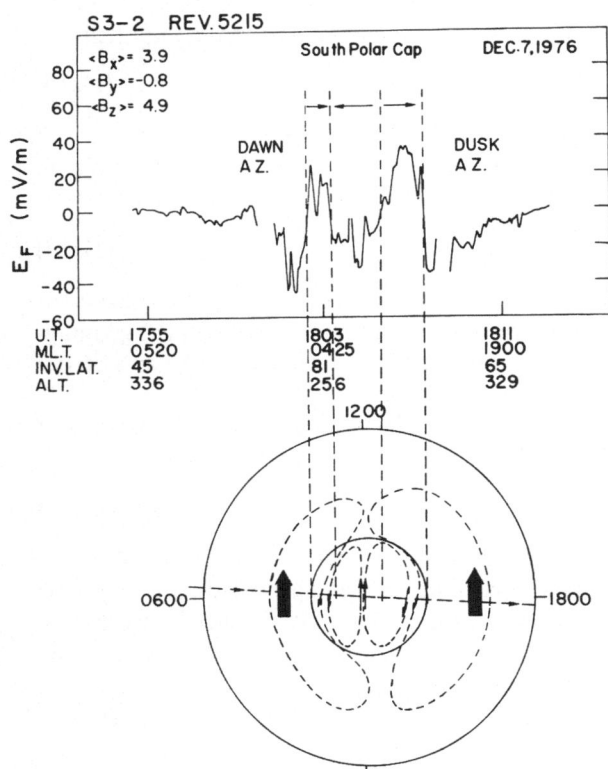

Fig. 11.5. Dawn–dusk component of the ionospheric electric field during a polar path of the satellite S3-2 on the southern hemisphere during a period of northward directed IMF (*upper panel*); and four-lobed convection pattern derived from these observations (*lower panel*) (Burke 1982)

al 1982), corresponding to an asymptotic electric field $E_y = E_{y0}\varrho_\infty$ in Eq. (11.9). In this case, the ionospheric electric field weakens toward the pole. A four-lobed pattern of the polar electric potential may even develop during a northward-directed IMF component (Fig. 11.5). The inner polar cap flow of this pattern is believed to be driven by the viscous-like interaction at the flanks of the magnetopause (Stern 1983).

11.2 Co-Rotation Field

Electric and magnetic fields depend on the coordinate system in which they are measured. The transformation from a rotating to a nonrotating frame of reference is given by

$$E' \simeq E + v \times B$$
$$B' \simeq B ,$$

$$(11.15)$$

where E', B' are the fields measured in the system rotating with (nonrelativistic) velocity v, and E, B are the fields in the nonrotating system. The plasma in the inner magnetosphere moves nearly rigidly with the earth due to viscous and neutral-ion drag forces with velocity v_0. Its zonal component is

$$v_{\lambda 0} = \Omega r \sin \theta ,$$

$$(11.16)$$

where $\Omega = 7.29 \times 10^{-5} \, \text{s}^{-1}$ is the angular frequency of the earth's rotation. If no electric field is measured in the rotating frame of reference ($E' = 0$), an electric field

$$E = - v_0 \times B_0$$

$$(11.17)$$

must exist in the nonrotating frame of reference in order to compensate for the Lorentz force $v_0 \times B_0$ which is set up by the plasma moving in the geomagnetic field B_0, and which otherwise would cause the plasma to drift away from the earth (e.g., Alfven and Fälthammar 1963).

In the case of a coaxial geomagnetic dipole and the field configuration of Eq. (11.3), the electric field [Eq. (11.17)] can be derived from the potential

$$\Phi_r = - \Phi_{r0}/L ,$$

$$(11.18)$$

with $\Phi_{r0} = \Omega a^2 B_{00} = 90 \, \text{kV}$, and L the shell parameter from Eq. (11.4).

The equipotential shells have torus-like shapes parallel to the magnetic field lines in Fig. 10.13b, and the maximum potential difference between pole and equator at ionospheric heights is 90 kV. An ionized particle within the inner magnetosphere possessing thermal energy can only drift orthogonal to E and B. The electric field seen by this particle is the superposition of the convection field and the co-rotation field. The $E \times B$ drift of the particle forces it to remain on an equipotential shell of the electric potential $\Phi = \Phi_c + \Phi_r$.

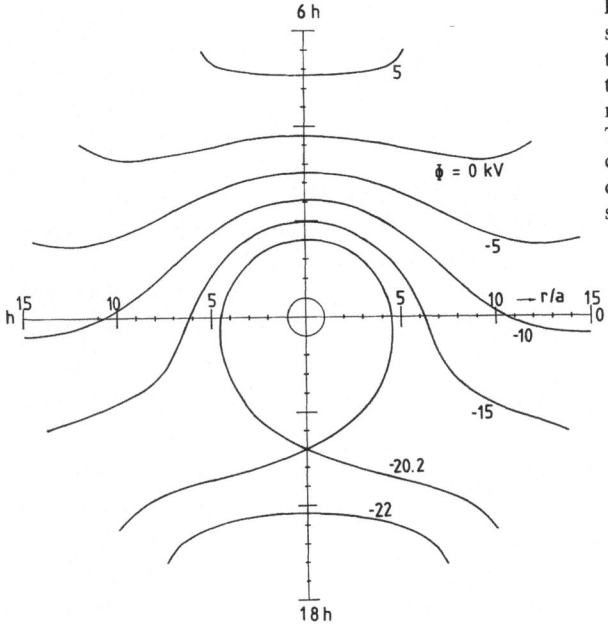

Fig. 11.6. Equipotential lines of sum of electric convection potential and co-rotation potential in the equatorial plane of the inner magnetosphere versus local time. The last closed equipotential line, called the plasmapause, is the outer boundary of the plasmasphere

Figure 11.6 shows the electric equipotential lines of this sum Φ within the equatorial plane of the inner magnetosphere. We now have two separate regions: one with closed equipotential lines near the earth where the co-rotation field dominates, and a second region of open equipotential lines where the electric convection field dominates. The region of closed equipotential shells is called the plasmasphere, and the last closed equipotential shell is called the plasmapause (e.g., Kivelson 1976). The thermal ions, mainly protons, are trapped within the plasmasphere. The ionosphere can provide thermal protons to the plasmasphere along the geomagnetic field lines so that a relatively high proton density is maintained ($n \gtrsim 10^8\,\mathrm{m}^{-3}$). These protons are created by charge exchange with oxygen ions below about 1000 km. Outside the plasmasphere, the protons can move freely along an open electric equipotential to the magnetopause where they may be permanently lost to the solar wind, so that the proton density in that region remains low ($n \lesssim 10^7\,\mathrm{m}^{-3}$). At the last closed electric equipotential shell – the plasmapause – a sudden drop in the proton density by several orders of magnitude occurs (e.g., Banks 1979).

It follows from Fig. 11.6 that a relative maximum of Φ exists at the point of maximum elongation of the plasmapause on the dusk side. The equation of the plasmapause can therefore be derived from the condition $\partial \Phi / \partial L = 0$ at $\tau = 270°$ as

$$\Phi_0 = (\Phi_{c0}/2)\,L^2 \sin \tau / L_b^2 - \Phi_{r0}/L \,, \tag{11.19}$$

with $\Phi_0 = -1.5\,\Phi_{c0}L_0^2/L_b^2$ the electric potential of the plasmapause, and $L_0 = (\Phi_{r0}L_b^2/\Phi_{c0})^{1/3}$ the maximum elongation of the plasmapause on the dusk side. With the numbers $\Phi_{r0} = 90\,\text{kV}$; $\Phi_{c0} = 30\,\text{kV}$; $L_b = 10$, we arrive at $\Phi_0 = -20.2\,\text{kV}$ and $L_0 = 6.7$. These values are typical for slightly disturbed conditions. L_0 decreases during disturbed conditions because Φ_{c0} increases and can reach values as low as $L_0 \simeq 4$.

The plasmapause closely follows any change in the convection field. The shape of the plasmapause may become quite complex during the recovery phase after storms (e.g., Chappel 1972; Spiro et al. 1981).

Rycroft and Thomas (1970) obtained an empirical relationship between the distance of the plasmapause on the night side L_{n0} ($\tau = 0$), and the magnetic index Kp:

$$L_{n0} = [8\,\Phi_{r0}L_b^2/(27\,\Phi_{c0})]^{1/3} = 5.64 - 0.78\,(Kp)^{1/2} \quad (Kp \lesssim 6)\,. \tag{11.20}$$

Inserting the same parametric values as above, we obtain $L_{n0} = 4.5$ and $Kp = 2.3$, indicating slightly disturbed conditions. For $Kp = 6$, we would obtain $L_0 = 5.6$, $L_{n0} = 3.7$ and $\Phi_{c0} = 52\,\text{kV}$. For $Kp = 0$, one has $L_0 = 8.6$ and $\Phi_{c0} = 15\,\text{kV}$. The uniform tail field from Eq. (11.9) is $E_y = 0.58\,\text{kV/a} = 91\,\mu\text{V/m}$ for $Kp = 2.3$ in reasonable agreement with observations (Kivelson 1976). For other empirical relations between L_0 and Kp, see Kivelson et al. (1980).

11.3 Polar Cap Field

Figure 10.9, showing the merging of the IMF lines with the geomagnetic field, suggests that an antisymmetric effect on the convection field should accompany the west-to-east directed IMF component B_y. In general, this component is directed from dawn to dusk ($B_y > 0$) during away polarity of the IMF and from dusk to dawn ($B_y < 0$) during toward polarity.

This antisymmetry has, in fact, been observed. Figure 11.7 schematically shows the electric convection field in the northern and in the southern hemisphere for both toward and away polarity. One sees a morning maximum within the polar caps in the northern hemisphere and an evening maximum in the southern hemisphere during away polarity (upper panel). This effect reverses during toward polarity (lower panel). It can be attributed to a zonally independent meridional electric field superposed on the symmetric convection field. This polar cap field is directed to the poles during away polarity and to the equator during toward polarity. Its maximum field strength is near $10°$ ($170°$) colatitude, and it disappears at middle and lower latitudes.

A polar cap electric field E', measured in the frame of reference of the rotating earth, is associated with differential rotation of the polar cap region of the magnetosphere. According to Eq. (11.16), the zonal velocity relative to the earth is

$$\Delta v_\lambda = v_\lambda - v_{\lambda0} = E_\theta'/(2\,|B_0|\cos\theta)\,. \tag{11.21}$$

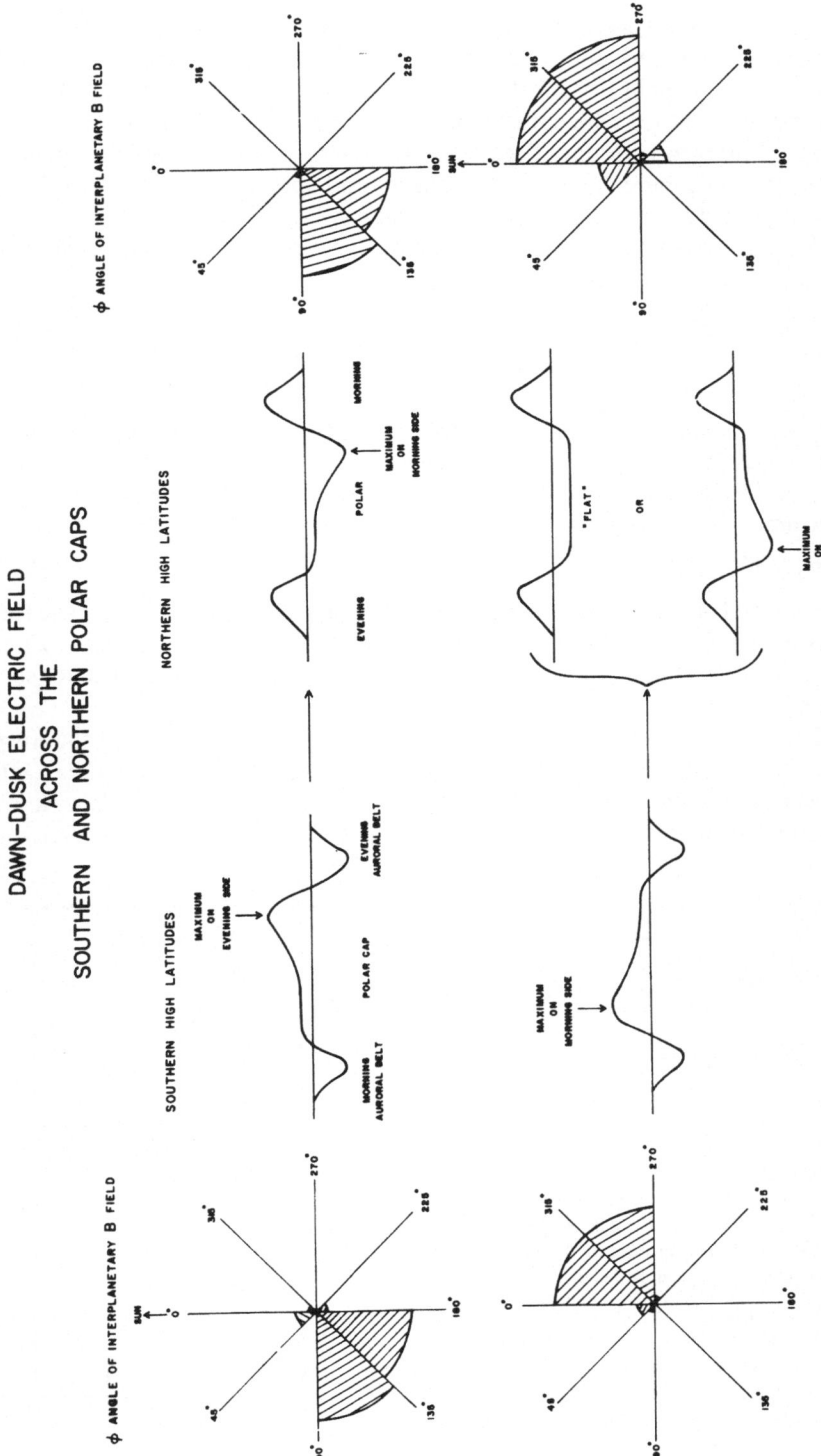

Fig. 11.7. Schematic view of superposition of convection field and polar cap field depending on the polarity of the interplanetary magnetic field. *Upper panel* away polarity of the IMF; *lower panel* toward polarity (Heppner 1972)

The polar cap magnetospheric plasma superrotates in the northern hemisphere ($\Delta v_\lambda > 0$) and rotates retrograde in the southern hemisphere ($\Delta v_\lambda < 0$) during toward polarity of the IMF. The rotation reverses during away polarity. Figure 11.8 schematically illustrates this effect. The magnetic tension along the merged IMF lines decelerates the corotating polar cap plasma in the northern hemisphere during away polarity and accelerates the plasma during toward polarity. The effect reverses in the southern hemisphere.

An analytic expression for the potential of this polar cap field, valid for the northern hemisphere, is

$$\Phi_p = \Phi_{p0} \begin{cases} 1 - (L_1 L_2)^{1/2}/L & \theta < \theta_1 \\ [1 - (L_2/L)^{1/2}]^2/[1 - (L_2/L_1)^{1/2}] & \text{for} \quad \theta_1 < \theta < \theta_2 \\ 0 & \theta_2 < \theta < 90° \end{cases} \quad (11.22)$$

with $|\Phi_{p0}| \simeq 3.4\,\text{kV}$; $\theta_1 \simeq 10°$; $\theta_2 \simeq 15°$; $L_i = 1/\sin^2 \theta_i$, and the shell parameter L from Eq. (11.4). Φ_{p0} is positive during toward polarity and negative during away polarity. The function (11.22) is antisymmetric with respect to the equator:

$$\Phi_p(180° - \theta) = -\Phi_p(\theta) . \quad (11.23)$$

Figure 11.9 shows the electric potential Φ_p and the meridional electric field E_θ as a function of colatitude θ during toward polarity. This polar cap field changes sign within about 1 h after a reversal of the IMF polarity (Clauer et al. 1983).

The downward mapping of magnetospheric electric fields into the lower atmosphere follows essentially the same line as discussed in Section 9.7. One expects observable effects on the ground only near the auroral ovals with vertical field strengths of typically ±3 V/m, perhaps increasing to 20 V/m and

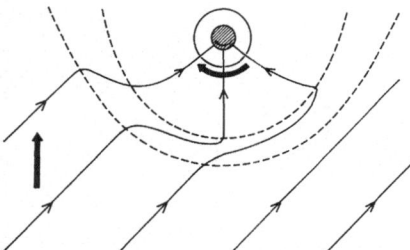

Fig. 11.8. Polar view of dayside reconnection between geomagnetic and interplanetary magnetic field lines during away polarity. *Thick arrows* indicate direction of motion of the newly reconnected field lines (Nishida 1979)

Fig. 11.9. a Electric potential, and **b** meridional ▶ electric field of polar cap potential at ionospheric heights on the northern hemisphere during toward polarity of the IMF. *Right abscissas* indicate stream function and zonal Hall current

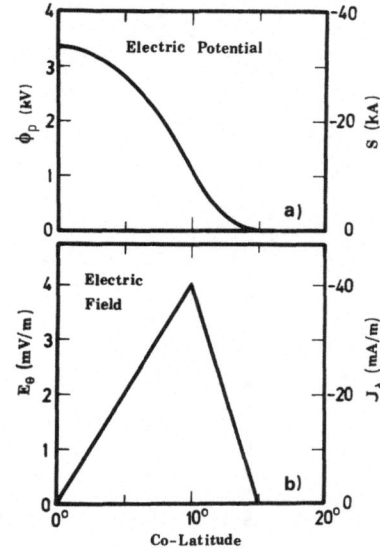

more during highly disturbed conditions (Park 1976). The mapping problem becomes more complicated for small-scale disturbances with horizontal scales smaller than 500 km, where nonpotential fields have to be taken into account (Boström and Fahleson 1977).

11.4 Electric Hall Current Systems

In the foregoing sections, we have discussed electric field configurations in the magnetosphere related to solar wind–magnetosphere interaction. It was shown that the field-aligned electric currents are connected within the dynamo region by ionospheric currents which we can determine from Eq. (11.1) provided U and Φ are known. These ionospheric currents cannot be measured directly, but must rather be derived from magnetic observations. A combination of simultaneous observations in space and on the ground leads, in principle, to a unique determination of the currents. Several difficulties, however, arise in practice: (a) Magnetic fields are integral effects of the currents. An inverse modeling is, therefore, not very accurate. (b) Space observations are relatively rare, and ground-based observations alone are not unique and can be interpreted as arising from a multitude of possible overhead current configurations.

Consider, for example, the simple current system (B) in Fig. 11.10, which consists of two infinitely extended vertical line currents (simulating the field-aligned currents), connected by a horizontal sheet with constant Pedersen conductivity (simulating the dynamo region). The azimuthal magnetic component at the lower end of one line current ($B2$ in Fig. 11.10) is

$$B_{\lambda 1} = \mu J_F/(4\pi\varrho) \,, \tag{11.24}$$

where J_F is the strength of the line current, and ϱ the horizontal distance from the line current [Eq. (14.25); note, however, that a line current extended into the half-space has a magnetic field, which is smaller by a factor of two com-

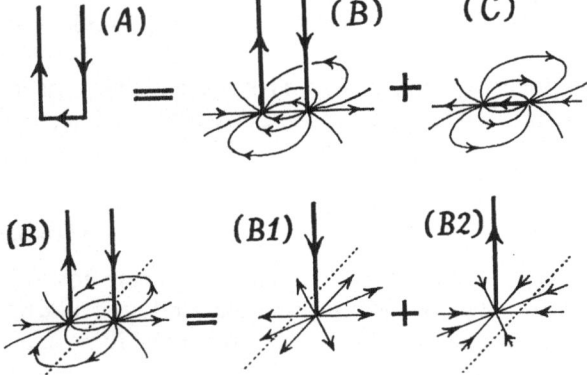

Fig. 11.10. Equivalence of different electric current systems as measured on the ground (Fukushima 1971)

pared with Eq. (14.25)]. The magnetic field due to the sheet current in the dynamo region also has only an azimuthal component:

$$B_{\lambda 2} \simeq -\mu J_p/2 .\tag{11.25}$$

Since the total current is divergence-free, the strength of the sheet current must be $J_p = J_F/(2\pi\varrho)$, so that $B_\lambda = 0$ on the ground. The magnetic effects of both currents thus cancel out. Ground-based observations of a current system like (A) in Fig. 11.10, therefore, can also be interpreted as arising from an equivalent horizontal sheet current system like (C) in Fig. 11.10. Horizontal sheet currents derived from ground-based magnetic records are called equivalent currents. System (C) in Fig. 11.10 is, therefore, the equivalent current of system (A).

The Pedersen currents connected with magnetospheric field-aligned currents thus have only a minor magnetic effect on the ground. The main contributors to magnetic fields on the ground are the Hall currents in the dynamo region and in the polar electrojets.

The Hall currents would flow entirely within the dynamo region, if the Hall conductivity were constant. Since $\nabla \cdot J_h = 0$ in this case, J_h can be derived from a stream function:

$$J_h = \nabla \times (\Psi \hat{r}) ,\tag{11.26}$$

with $\Psi = -\Sigma_h \Phi$. It follows that the equipotential lines in Fig. 11.2 are also stream lines of the Hall currents.

The Hall current of the convection field in Fig. 11.2 consists of two vortices in each hemisphere flowing in clockwise direction on the morning side and in couter-clockwise direction in the evening side on the northern hemisphere as seen from above. The total current within one vortex (with $\Sigma_h \simeq 10$ S) is

$$\Phi_{\max} = \Sigma_h \Phi_{c0}/2 \simeq 150 \text{ kA} .\tag{11.27}$$

Equivalent currents of similar configuration with current strengths of this order of magnitude have, indeed, been derived from geomagnetic observations in polar and subpolar regions. They are called S_q^p currents during quiet conditions and $DP2$ disturbances during disturbed conditions (Nishida and Kokubun 1971).

The Hall currents of the polar cap field in Fig. 11.9 encircle the poles in clockwise direction, as seen from above, during toward polarity, and in counter-clockwise direction during away polarity of the IMF. The total current in one hemisphere is

$$\Psi_{\max} = \Sigma_h \Phi_{p0} \simeq 34 \text{ kA} .\tag{11.28}$$

The magnetic effect of this circumpolar Hall current during toward polarity is a downward-directed magnetic field at the poles and an equatorward directed field just below the maximum flows near 10° and 170° colatitude. The directions reverse sign for away polarity of the IMF. It was these polar geomagnetic disturbances, which were observed to depend on the azimuthal component of the IMF and thus, in general, on its polarity, which led to the discovery of the polar cap field (Svalgaard 1969; Mansurov 1969). Since geomagnetic measurements near the geomagnetic north pole (Thule, Greenland) date back to 1926, it is possible to evaluate the IMF polarity continuously from this time on (Svalgaard 1973).

The electric conductivity in the dynamo region, of course, varies with time of day, season, and latitude. The Hall current of the polar cap field, therefore, flows mostly during sunlit hours and closes via Birkeland currents. Field-aligned currents associated with the polar cap field flow poleward of the region I currents in the noon sector, changing their direction with changing y-component of the IMF (Saflekos et al. 1982).

If the spatially and temporally varying elements of the electric conductivity tensor in the dynamo region are assumed to be known, it is possible to decouple the ionospheric currents from the field-aligned currents from ground-based magnetic records alone (Kamide et al. 1981). Other methods to separate these two current systems are due to Baumjohann et al. (1980) and Zanetti et al. (1983).

11.5 Polar Electrojets

Solar wind–magnetospheric interaction causes a never-ceasing, but highly fluctuating energy input into the auroral ovals, which leads to an almost permanent enhancement of the electric conductivity in these regions. The just as permanently existing electric convection field drives a band of electric currents in the auroral ovals – the polar electrojets. The quasi-steady-state behavior of these convection electrojets is shown in Fig. 10.12a. They are part of the DP2 disturbance discussed in Section 11.4.

Polar electrojets are very complex phenomena depending on the time history of the electric convection field as well as on the real configuration of the ionospheric conductivity. Characteristic regular features can be derived, however, if one averages over a large number of individual storms.

We shall model the typical behavior of convection electrojets in the following. For this purpose, we need to modify our convection field from Section 11.1. We introduce a zone of enhanced conductivity, $(\Sigma_p + \Delta\Sigma_p, \Sigma_h + \Delta\Sigma_h)$, in the dynamo region situated between colatitudes (θ_b, θ_s). This zone simulates the auroral oval in the northern hemisphere. A corresponding zone lies in the southern hemisphere. Since the convection potential is symmetric with respect to the equator, we consider only the northern hemisphere. The conductivities Σ_p and Σ_h remain constant outside the auroral zones. The electric potential in the polar cap and, therefore, the asymptotic electric field E_y in the magnetospheric tail remain the same as in Eq. (11.9). However, the potential in the auroral zone $(\theta_b < \theta < \theta_s)$ and at lower latitudes $(\theta_s < \theta < 90°)$ must be modified:

$$\Phi^* = -i(\Phi_{c0}/2)\exp(i\tau)\begin{cases}(L_b/L)^{1/2} & L_b < L \\ (L/L_b)^{s/2} & \text{for} \quad L_s < L < L_b \quad (11.29)\\ (L/L_s)^{q/2}[L_s/L_b]^{s/2} & 1 < L < L_s\end{cases}$$

where L is again the shell parameter and $L_i = 1/\sin^2\theta_i$. We use here for convenience a complex representation of Φ. The physically relevant solution is now $\Phi = \text{Real}\,(\Phi^*)$. The parameter q is the same as in Eq. (11.10), and s will be determined below.

Φ^* is continuous at the boundaries L_b and L_s. Furthermore, the meridional electric current must be continuous at the low latitude border of the auroral zone at θ_s. This requires that

$$(\Sigma_p E_\theta + \Sigma_h E_\lambda)\Big|_{\theta_s+0} = [(\Sigma_p + \Delta\Sigma_p)E_\theta + (\Sigma_h + \Delta\Sigma_h)E_\lambda]\Big|_{\theta_s-0}, \tag{11.30}$$

from which follows the parameter s in Eq. (11.29) as

$$s = \alpha + i\beta = q\Sigma_p/(\Sigma_p + \Delta\Sigma_p) + i\Delta\Sigma_h/[(\Sigma_p + \Delta\Sigma_p)\cos\theta_s]. \tag{11.31}$$

Similarly, the field-aligned current at θ_b can be derived from the difference of the meridional currents on the polar side and on the equatorward side of the electric field reversal at θ_b:

$$J_F^* = (J_\theta\Big|_{\theta_b-0} - J_\theta\Big|_{\theta_b+0})/\sin I = i\,\Phi_{c0}(L_b/L_s)^{1/2}\cot\theta_b[\Sigma_p(q+1)$$
$$- i\Delta\Sigma_h(1/\cos\theta_b - 1/\cos\theta_s)]\exp(i\tau)/(2a\sin I). \tag{11.32}$$

In this case, the Hall current is not source free within the dynamo region, and a small part of the Hall current closes via Birkeland currents. The magnitude of the Birkeland current in Eq. (11.32) depends only weakly on the zone of enhanced electric conductivity in the auroral zone, as a comparison with Eq. (11.13) indicates.

Finally, the polar electrojets J_E are the excess currents in the auroral zone. Their dominant longitudinal component is

$$J_{E\lambda}^* = \Delta\Sigma_p E_\lambda - \Delta\Sigma_h E_\theta =$$
$$- \Phi_{c0}(L/L_b)^{s/2}(\Delta\Sigma_p - is\Delta\Sigma_h\cos\theta)\exp(i\tau)/(2a\sin\theta) \tag{11.33}$$

for $(\theta_b < \theta < \theta_s)$.

We use the following numbers for our simulation: $\Sigma_p = \Sigma_h = \Delta\Sigma_p = \Delta\Sigma_p = 10\,\text{S}$; $\theta_b = 18.4°$ ($L_b = 10$); $\theta_s = 23.4°$ ($L_s = 6.3$); $\Phi_{c0} = 30\,\text{kV}$; $q = 4$. In this case, $\alpha = 2$ and $\beta = 0.54$ from Eq. (11.31). The strip of enhanced electric conductivity in the auroral zone causes the meridional electric field to decrease by a factor $|s|/q \simeq 0.52$ of its value without enhanced conductivity in the auroral zone. The meridional field at lower latitudes, however, increases by a factor $[L_b/L_s]^{(q-\alpha)/2} \simeq 1.58$ compared with the undisturbed case ($\Delta\Sigma_p = \Delta\Sigma_h = 0$) [see Eq. (11.11)]. This decrease of the meridional electric field in the auroral zone and its corresponding increase at lower latitudes can be attributed to a secondary electric field resulting from electric polarization charges at the borders of the auroral zone so that flow continuity is maintained across the boundaries of discontinuous conductivity (Böstrom 1964). Moreover, the time of zero E_θ shifts from $\tau = 12 - \tau_1$ at L_b to $\tau = 12 - \tau_1 - \Delta\tau$ at L_s within the auroral zone where $\tau_1 = \arctan(\beta/\alpha) \simeq 61$ min, and $\Delta\tau = (\beta/2) \ln[L_s/L_b] \simeq 29$ min. The line where $E_\theta = 0$ divides the westward directed from the eastward directed $E \times B$ plasma flow at ionospheric F-layer heights. This line, called the Harang discontinuity, can be observed mainly during night time hours. It typically changes from about 2230 LT at $\theta_b \simeq 70°$ to about 2330 LT at $\theta_s \simeq 65°$ (Maynard 1974), in general agreement with our calculations.

The phase shift of Φ^* in Eq. (11.29) at lower latitudes ($L < L_s$) also shifts the location of maximum elongation of the plasmapause from 1800 LT to $1800 + \Delta\tau \simeq 1830$ LT. This phase shift toward the evening sector is a common phenomenon of the steady-state behavior of the plasmapause.

Two polar electrojets flow in our model [Eq. (11.33)]: one toward the west with a maximum near 0400 LT, the other one toward the east, maximizing near 1600 LT. This is the idealized distribution derived from averaging over a large number of geomagnetic disturbances (see Fig. 10.12a). The maximum current strength is of the order of

$$\overline{|J_{\lambda E}|} a(\theta_s - \theta_b) \simeq 100 \text{ kA} . \tag{11.34}$$

The sheet current density of $|J_{\lambda E}| \simeq 0.2$ A/m corresponds to a horizontal magnetic field of the order of $\Delta H \simeq 2\mu|J_{\lambda E}|/3 \simeq 170$ nT [see Eq. (8.2)] below the center of one polar electrojet. This is the order of magnitude of magnetic variations observed during moderately disturbed conditions below a polar electrojet (e.g., Untiedt et al. 1978).

The main contribution to the polar electrojet is a Hall current, (in our example: 75%), generated by the meridional component of the secondary electric field within the auroral oval. This secondary electric field acts in the same manner as in the case of the equatorial electrojet to increase the effective electric conductivity [Eq. (9.26)]. Due to the inhomogeneous Hall conductivity in the dynamo region, field-aligned Hall currents close within the magnetosphere (Fig. 10.12a).

The field and current configurations discussed in this chapter, of course, describe only gross steady-state features of a much more complicated phenomenon.

11.6 Ring Current

High energy ionized particles may be trapped within the inner magnetosphere if their pitch angles fulfill certain conditions. A particle's pitch angle α is the angle between the particle velocity vector and the geomagnetic field line along which the particle gyrates. Particles gyrating along a dipole line (Fig. 10.8) are reflected at mirror points where their pitch angles become $\pi/2$. The pitch angle at the equator α_e is related to the geomagnetic dipole field at the mirror point B_0 and at the equator B_e by (e.g., Roederer 1970)

$$\sin^2\alpha_e = B_e/B_0 = \sin^6\theta_0/(1+3\cos^2\theta_0)^{1/2}, \tag{11.35}$$

where θ_0 is the colatitude of the mirror point. Here, it is assumed that the particles do not encounter collisions and that their velocities are nonrelativistic.

In the presence of a nonelectric external force F, the particles also drift orthogonal to F and B with a velocity (Schulz and Lanzerotti 1974)

$$v_D = F \times B_0/(QB_0^2). \tag{11.36}$$

Since the drift velocity depends on the electric charge Q of the particles, positive ions drift westward and electrons drift eastward in the geomagnetic dipole field if F is directed outward (see Fig. 10.8).

The gradient of the geomagnetic dipole field is such an external force. Its strength at the equator is

$$F_t = 3W_t/r, \tag{11.37}$$

where $W_t = mv_t^2/2$ is the transverse kinetic energy of the particles and v_t is their velocity transverse to B_0.

A centrifugal force also acts on the particles that move along the curved field lines. This force generates a curvature drift and has the magnitude

$$F_p = 2W_p/r_c, \tag{11.38}$$

where $W_p = mv_p^2/2$ is the kinetic energy of the particles parallel to the geomagnetic field, v_p is the parallel velocity, and r_c is the radius of curvature of the field line. The dipole field at the equator has $r_c = r/3$.

The gradient drift and the curvature drift point in the same direction and always appear together. The combined gradient-curvature drift at the equator, directed to the west for positively charged particles, is

$$v_D = -12\pi r^2 W(1+\cos^2\alpha_e)/(\mu QM) \tag{11.39}$$

where $W = W_t + W_p$ is the total kinetic energy of the particles and $M = 8 \times 10^{22}\,\mathrm{Am}^2$ is the magnetic moment of the geomagnetic dipole field. Protons with energies of 100 keV need about 15 s to bounce back and forth between mirror points and about 2 h to drift around the earth at $L = 4$.

Protons with energies >30 MeV populate a relatively stable inner radiation belt (van Allen belt) which is centered near 4000 km altitude. They are magnetically undetectable because of their low density. The outer radiation belt is filled mainly with protons and electrons of lower energies. Protons, possibly also O^+ and He^+ ions, in the energy range below about 200 keV are the main contributors to an electric ring current centered near 6.5 earth radii during quiet conditions and near 4 earth radii or less during disturbed conditions. The ring current can be magnetically detected on the ground (Fig. 11.11). This disturbance is called Dst.

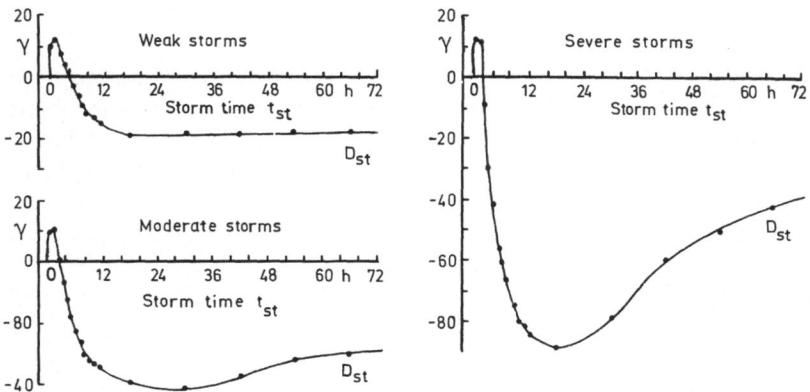

Fig. 11.11. Representative magnetic deviations of ΔH during geomagnetic storms, observed at low latitudes. (Sugiura and Chapman 1960) (Note that $1 \gamma = 1$ nT)

A single positive ion with pitch angle $\pi/2$ at the equator drifting around the earth in westward direction with velocity v_D acts like a ring current of strength $J = Qv_D/(2\pi r)$. The magnetic field of such a ring current at its center is (Fig. 14.3b)

$$\Delta H_1 = \mu J/(2r) = -3\,W/M\,. \tag{11.40}$$

The minus sign indicates that ΔH_1 is opposite to the dipole field. It is southward-directed and weakens the geomagnetic field on earth at low latitudes.

Particles also gyrate around the geomagnetic field lines. The magnetic effect of this gyration is to weaken the local value of the primary field. Gyrating ions therefore behave like a diamagnetic medium. At large distances from the gyration center, this magnetic field behaves like a dipole field with a magnetic moment W/B_0. Its magnetic field at the earth's center is

$$\Delta H_2 = \mu W/(4\pi r^3 B_0) = W/M\,. \tag{11.41}$$

In this case, the field adds to the dipole field, and the total field perturbation on the earth due to a single particle becomes

$$\Delta H = \Delta H_1 + \Delta H_2 = -2\,W/M\,. \tag{11.42}$$

For a total of N ring current particles, the component of a geomagnetic disturbance field at the equator on the ground is

$$\Delta H = -8NW/(3M)\,. \tag{11.43}$$

Here, we have taken into account the induction effect within the earth's interior which increases the field by about 33% [Eq. (8.2)]. A 100 keV proton has an energy of $W = 10^5\,eV = 1.6 \times 10^{-14}\,J$. A total of $N \simeq 1 \times 10^{29}$ of 100 keV protons is needed to generate a field of $\Delta H = 50$ nT on the ground. If these protons are stored in a torus of diameter of $4a$ located at $\bar{r} = 4a$ distance from the earth, the average proton density becomes $n \simeq N/(32\pi^2 a^3) \simeq$

$1 \times 10^6 \, \mathrm{m}^{-3}$. This corresponds to an equivalent ring current at $\bar{r} = 4a$ of strength $J = 3\bar{r}\Delta H/(2\mu) \simeq 1.5 \, \mathrm{MA}$.

Concentrations of $10 - 200 \, \mathrm{keV}$ particles of this order have indeed been detected near $L = 3 - 5$ (e.g., Rees and Roble 1975), and negative magnetic field deviations of 50 nT and more have been measured on the ground during severe magnetic storms (Fig. 11.11). The recovery to normal conditions takes several days. Such disturbances are measured by low latitude magnetometers; and these records are used to compute an index of the strength of the symmetric component of the ring current — the *Dst* index.

More sophisticated models of the ring current take into account the real particle distribution in space and energy and the mutual interaction of the particles (e.g., Sckopke 1972; Siscoe 1979). The magnetic field produced by all particles may become of the same order of magnitude as the undisturbed dipole field in the environment of the ring current. Therefore, the geomagnetic field from which the current was calculated changes and, in turn, influences the ring current.

The generation of the ring current and its temporal variation are not well understood because the composition and thus the source is still a matter of debate. The dawn–dusk electric convection field which moves the low latitude magnetospheric plasma toward the earth, but also acceleration of ionospheric ions by electric fields parallel to auroral geomagnetic field lines have been discussed as possible sources. The decay of the ring current during its recovery phase is probably due to scattering of the particles at the mirror points, or to charge exchange with the cold plasma in the plasmapause region. For more details see Williams (1980, 1983).

11.7 Magnetospheric Substorm

The physical processes involved in the phenomenon of the magnetospheric substorm remain a controversal matters. We have already mentioned in Section 10.4 (Fig. 10.14) two basic mechanisms which may explain the physical processes involved — the current disruption model and the reconnection model. We discuss in more detail in this section the current disruption model based on recent studies by Rostoker (1983). This model is easier to conceptualize than the reconnection model and fits better into the general scheme of a driven process outlined in Sections 10.3 and 10.4. The reader should, however, be aware of the existence of other mechanisms which may either simultaneously act or even replace the current disruption model.

Figure 11.12 shows a prototype of the magnetogram of a magnetospheric substorm observed on the ground according to Rostoker (1983), (rarely realized in this idealized form in nature). After a longer-lasting period of a northward-directed IMF and the magnetosphere in its ground state, the IMF turns abruptly southward (lower panel), and the magnetospheric plasma gains ex-

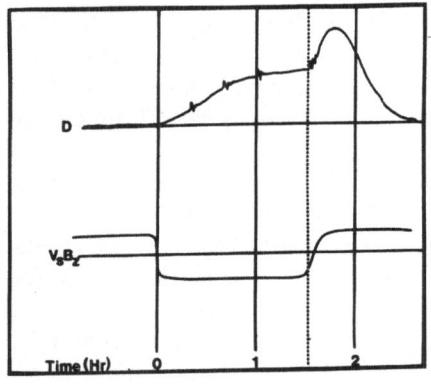

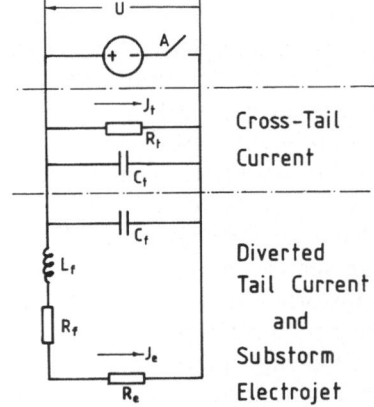

Fig. 11.12   **Fig. 11.13**

Fig. 11.12. Schematic illustration of time development of an isolated substorm initiated by a southward turning IMF. Lower panel shows the behavior of the IMF. Upper panel shows the corresponding magnetic variation below the substorm electrojet. At the time of southward turning of the IMF (*vertical line* at time 0), the substorm electrojet starts to flow (onset of the substorm) and increases in strength (growth phase) till a steady state is reached after about 1 h. The growth is interrupted by sporadic geomagnetic pulsations of the pi2 type (see Sect. 13.1). After northward turning of the IMF (at the *vertical dotted line*), the substorm electrojet increases explosively (expansion phase) and then returns to its ground state (recovery phase) (Rostoker 1983)

Fig. 11.13. Lumped circuit of substorm electrojet and tail current during an idealized substorm event

cess power from the solar wind. The rate at which energy from the solar wind is coupled into the magnetosphere depends on the product of solar wind velocity and southward component of the IMF. This product is also a measure of the electric convection field [Eq. (10.8)]. The plasma sheet in the tail begins to thin (Fig. 10.14b), and the tail plasma is heated. Part of this energized plasma penetrates into the auroral ovals. The auroral ovals expand and the electric conductivity in this area increases. The sunward streaming plasma in the tail fills the ring current. At the same time, the electric convection field increases, the DP2 system amplifies, and the field-aligned currents in region I are enhanced [Eq. (11.32)]. Moreover, the cross-tail current (Fig. 10.14b) is disrupted and diverted to the auroral ionosphere along magnetic field lines generating the substorm electrojet (Fig. 10.12b). (We again consider only the northern hemisphere.) The geomagnetic disturbance in the environment of the substorm electrojet is a measure of its strength (Fig. 11.12; upper panel).

The amplitude increase after onset of the substorm is called the growth phase. If a time of a few hours elapses without changing IMF, the growth phase reaches a steady state. The geomagnetic pulsation activity during the growth phase indicated in Fig. 11.12 will be discussed in Section 13.1.

The energy stored in the tail is released when the IMF turns northward again. The eruptive increase of the geomagnetic disturbance after northward turning of the IMF (the time at the vertical dotted line in Fig. 11.12) is called the expansion phase. After reaching a maximum within a fraction of an hour, the magnetic deflection returns to its ground state (recovery phase). The DP2 disturbance is weakened with the onset of the expansion phase (Kamide and Baumjohann 1984).

We want to simulate the temporal variation of the system consisting of the cross-tail current and the diverted tail current system including the substorm electrojet in a simple model. Figure 11.13 shows the lumped circuit representing such an event. R_e is the resistance of the ionospheric section where the substorm electrojets of both hemispheres flow in parallel connection. It may be a complex impedance including the region II field-aligned currents and the partial ring current (Siscoe 1982). R_f, L_f, and C_f are resistance, inductance, and capacitance of the field-aligned section of the substorm current. The elements R_t and C_t are the corresponding resistance and capacitance of the cross-tail current. A possible inductance of the tail current is neglected for convenience. The system is driven by a voltage source U simulating the step-like increase of the electric convection field after southward turning of the IMF.

For an estimate of the elements in the electrojet circuit, we consider the field-aligned section of this system as consisting of two adjacent current sheets each of height h and length d. They are separated by the width b. Then, the inductance of this system is related to the magnetic energy density W_m as (Magid 1972; Rostoker and Böstrom 1976)

$$W_m = 1/(2\mu) \int_V B^2 dV \simeq B^2 bhd/(2\mu) \simeq (1/2) L_f J_f^2 , \qquad (11.44)$$

where J_f is the strength of the electric current in the sheets, $V = hbd$ is the volume of the system, and B is the magnetic field generated by this current. It is related to J_f by $B \simeq \mu J_f/d$ so that

$$L_f \simeq \mu bh/d . \qquad (11.45)$$

The capacitance of the two sheets is estimated to be proportional to the kinetic energy of the plasma in the volume V:

$$W_k = \int_V \varrho v^2/2 \, dV \simeq \varrho E^2 bhd/(2B^2) \simeq (1/2) C_f \Phi^2 , \qquad (11.46)$$

where $v \simeq E/B$ is the plasma velocity, ϱ is the plasma density and $\Phi \simeq Eb$ is the electric voltage between the sheets. Thus,

$$C_f \simeq \varrho hd/(bB^2) . \qquad (11.47)$$

With the numbers $b \simeq h \simeq 3$ to $10a$; $d \simeq 15a$; $\varrho \simeq 1 \times 10^{-21} \, \mathrm{kg/m^3}$; $B \simeq 10$ to 30 nT, one arrives at $L_f \simeq 5$ to 50 H and $C_f \simeq 0.1$ to 1 kF.

The capacitance of the cross-tail current regime may be estimated from Eqs. (10.12) and (11.46) as

$$C_t \simeq 2 W_{\mathrm{tail}}/\Phi^2 \simeq 0.1 \text{ to } 1 \text{ MF} \qquad (11.48)$$

(with $\Phi \approx 50$ to $150\,\text{kV}$). The resistances of the cross-tail regime and the electrojet regime will be estimated below.

Southward turning of the IMF is simulated by switching on the voltage to a constant U_0 at time $t = 0$ (closing gate A in Fig. 11.13). The response of the electrojet current is

$$J_e = J_{e0}[1 - \exp(-\gamma t)] \quad (t \geq 0), \tag{11.49}$$

where $J_{e0} = U_0/(R_f + R_e)$ and $\gamma = (R_f + R_e)/L_f$. This corresponds to the growth phase of the substorm. $J_{e0}/2$ is the maximum steady-state current of the substorm electrojet within one hemisphere.

The condensers C_f and C_t are in parallel connection and are immediately stored with a charge of

$$Q = U_0(C_f + C_t) \tag{11.50}$$

after switching on the voltage. This simulates the storage of solar wind energy within the magnetosphere during a substorm.

If the voltage is switched off again after the growth phase has reached a steady state, the stored energy in the condensers and inductor is unloaded and dissipated. The response of the system is via its eigenmodes. The currents in the electrojet circuit and in the tail decay as

$$J_e = J_{e0}[(\beta - 2R/L)\exp(-\alpha t) - (\alpha - 2R/L)\exp(-\beta t)]/(\beta - \alpha)$$
$$J_t = J_{t0}[(\alpha L/R - 1)(\beta - 2R/L)\exp(-\alpha t) - (\beta L/R - 1) \tag{11.51}$$
$$\cdot (\alpha - 2R/L)\exp(-\beta t)]/(\beta - \alpha)$$

with

$$\left.\begin{array}{r}\beta\\\alpha\end{array}\right\} = \{(GL + RC) \pm [(GL - RC)^2 - 4LC]^{1/2}\}/(2LC)$$

and $C = C_t + C_f \approx C_t$; $L = L_f$; $G = 1/R_t$; $R = R_f + R_e$; $J_{t0} = GU_0$ [see for comparison, Fig. 6.5 and Eq. (6.10)].

These currents become particularly simple if we assume $\alpha = \beta$, which yields

$$J_e = J_{e0}\exp(-\alpha t)[1 - (3RC - GL)t/(2LC)]$$
$$J_t = J_{t0}\exp(-\alpha t)[1 - (1 - \alpha L/R)(\alpha - 2R/L)t] \tag{11.52}$$

with $\alpha = (GL + RC)/(2LC)$. This decay can be considered as the recovery phase.

If one selects the numbers $U_0 = 60\,\text{kV}$; $L_f = 30\,\text{H}$; $C_f = 300\,\text{F}$; $R = 15\,\text{m}\Omega$; $C_t = 130\,\text{kF}$; $G = 200\,\text{S}$ (or $R_t = 1/G = 5\,\text{m}\Omega$), one arrives at $J_{e0} = 4\,\text{MA}$; $J_{t0} = 12\,\text{MA}$; $\gamma = 5 \times 10^{-4}\,\text{s}^{-1}$; (or $\tau_1 = 1/\gamma = 2000\,\text{s}$); $\alpha = 1 \times 10^{-3}\,\text{s}^{-1}$ (or $\tau_2 = 1/\alpha = 1000\,\text{s}$); $P_{\text{mag}} = (J_{e0} + J_{t0})U_0 = 1\,\text{TW}$; $Q = 8\,\text{GC}$.

The characteristic time of the growth phase τ_1 and the characteristic time of the recovery phase τ_2 are consistent with observations (Fig. 11.12). The power P_{mag} added to the magnetosphere corresponds to medium size sub-storms [Eq. (10.11)].

Growth and recovery phases are driven processes, according to this picture. They can therefore be simulated by a simple linear model. However, the expansion phase cannot be treated in such a manner. This phase is probably a nonlinear process related to the reconfiguration of the tail structure after northward turning of the IMF. If, for instance, the conductance G of the cross-tail current would suddenly decrease after northward turning of the IMF, a relatively larger part of the total unloading current would flow via the electrojet circuit, as the expansion phase suggests. It is during the expansion phase that reconnection processes, neglected so far, may play a leading role.

An individual substorm is, or course, much more complex. The IMF may turn several times within time scales small compared to the characteristic time of the system. The circuit elements in Fig. 11.13 certainly vary with time, and the lumped circuit is a gross oversimplification. Detailed calculations simulating particular substorm events have been undertaken by Harel et al. (1981) and Wolf et al. (1982).

12 Theory of Wave Propagation Within the Magnetosphere

Electromagnetic and hydromagnetic waves can be excited within the magnetosphere either by solar wind–magnetospheric interaction or by internal instabilities. Electromagnetic pulses generated by lightning events can propagate through the magnetosphere along the geomagnetic field lines and can be observed on the ground and within the magnetosphere. In the next two chapters, we present a short introduction into the theory of wave propagation through a magnetoactive plasma, followed by a brief discussion of observed wave phenomena in the magnetosphere.

12.1 Ray Theory

Plasma waves are waves involving electric and magnetic fields, ions and electrons. We call them electromagnetic waves if their angular frequencies ω are comparable with or larger than the gyrofrequency of the ions [ω_{Hi} in Eq. (2.12)]. They are called hydromagnetic waves if their angular frequencies are much smaller than ω_{Hi}.

In the following, we deal with harmonic waves of frequencies $f = \omega/(2\pi) < 100$ kHz propagating in a cold plasma in which the plasma pressure can be neglected. The simplest configuration is wave propagation within a homogeneous, infinitely extended plasma in the presence of a homogeneous external magnetic field $\boldsymbol{B}_0$. The electric and magnetic components of a plane wave propagating in the z-direction of a Cartesian coordinate system are proportional to

$$\exp(iknz - i\omega t),\tag{12.1}$$

with $k = \omega/c$ the wave number in vacuo, c the speed of light, and n the refractive index of the medium. The refractive index determines the phase velocity v_p and the group velocity v_g of the waves:

$$v_p = c/n; \quad v_g = c/(n + \omega \partial n/\partial \omega).\tag{12.2}$$

The group velocity describes the velocity of pulse propagation. The magnetic field vector of the wave oscillates in the wave plane orthogonal to the propagation direction (in our case in the (x, y)-plane). The ratio of the magnetic field components determines the wave polarization:

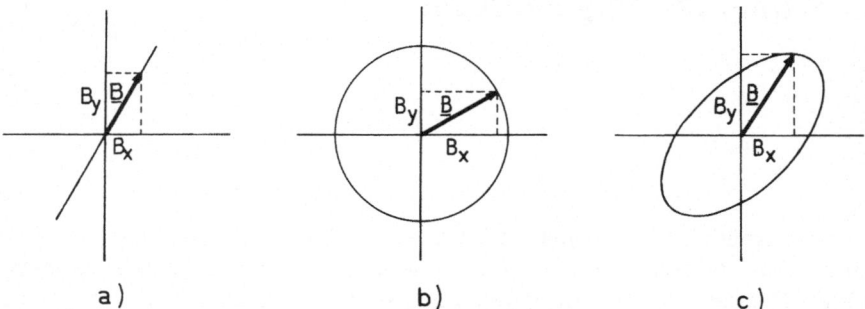

Fig. 12.1a – c. Polarization of electromagnetic waves perpendicular to wave propagation. **a** linear polarization; **b** circular polarization; **c** elliptical polarization

$$B_y/B_x = A \exp(\pm i\phi) . \qquad (12.3)$$

If this ratio is real ($\phi = 0$, π), the wave is linearly polarized (see Fig. 12.1a), and the wave vector oscillates along a straight line. If $A = 1$ and $\phi = \pm \pi/2$, the wave is circularly polarized (Fig. 12.1b), and the wave vector rotates either clockwise [CW; upper sign in Eq. (12.3)] or counterclockwise [CCW; lower sign in Eq. (12.3)]. The general case is an elliptically polarized wave, either CW or CCW rotating (Fig. 12.1c).

A relationship between the horizontal components of the electric and magnetic field follows from Maxwell's equations:

$$E_x/E_y = -B_y/B_x , \qquad (12.4)$$

which means that the electric field component in the wave plane is always perpendicular to the magnetic field vector.

The wave field forces the electrons and ions in the plasma to oscillate. This oscillation generates a secondary electromagnetic wave which is superposed onto the primary wave. The reaction of the plasma to the wave field is described by the refractive index n. If the motion of the electrons and the main ion constituent in the wave field is determined in a simplified way via Ohm's law neglecting collisions with neutral particles, one may derive a dispersion relationship from Maxwell's equations connecting the refractive index with the wave frequency. In the case of wave propagation in the direction of the magnetic field [longitudinal propagation; $B_0 = (0, 0, B_0)$], this relation is (e.g., Ratcliffe 1972)

$$n^2 = 1 - X/[(1 \pm Y_e)(1 \mp Y_i)] , \qquad (12.5)$$

with

$$X = X_e + X_i \simeq X_e; \quad X_e = (\omega_e/\omega)^2; \quad X_i = (\omega_i/\omega)^2;$$

$$Y_e = \omega_{He}/\omega; \quad Y_i = \omega_{Hi}/\omega$$

[ω_e and ω_i from Eq. (2.11); ω_{He} and ω_{Hi} from (2.12)]. Furthermore, the waves are circularly polarized:

$$B_y/B_x = \pm i . \qquad (12.6)$$

A positive ion gyrating in a magnetic field rotates counterclockwise (CCW) if one looks in the direction of the magnetic field (Fig. 2.4). A wave with the same sense of polarization is called an ion wave or i-wave (referred to by the index "i"). A wave with the same sense of polarization as that of a clockwise (CW) gyrating electron is called an electron wave or an e-wave (index "e"). The relation between i-waves and e-waves, on the one hand, and CW and CCW polarization, on the other hand, therefore, depends on whether the magnetic field is directed toward or away from the observer.

Figure 12.2 (left) shows n^2 plotted versus X [or $f = \omega/(2\pi)$] for longitudinal propagation. The refractive index of both waves starts at $n = 1$ for very large frequencies ($X \rightarrow 0$). For small departures from exact longitudinal propagation, the i-trace separates into two traces, so that three zeros of n^2 exist at $X = 1$ and $X = (1 \pm Y_e)(1 \mp Y_i)$ or at frequencies

$$f = \begin{cases} f_0 = (f_e^2 + f_i^2)^{1/2} \simeq f_e \\ \pm (f_{He} - f_{Hi})/2 + [(f_{He} + f_{Hi})^2/4 + f_0^2]^{1/2}. \end{cases} \tag{12.7}$$

n^2 becomes infinite at $Y_e = 1 \, (f = f_{He})$ and at $Y_i = 1 \, (f = f_{Hi})$. Both waves reach the same $n = c/v_a$ at zero frequency ($X \rightarrow \infty$), where

$$v_a = c[1 + X/(Y_e Y_i)]^{-1/2} \simeq B_0/(\mu \varrho_i)^{1/2} \tag{12.8}$$

is the Alfven velocity [$\varrho_i = n_i(m_i + m_e) \simeq n_i m_i$ is the ion density]. The polarization of the waves is indicated in Fig. 12.2 by the symbols "i" and "e", respectively.

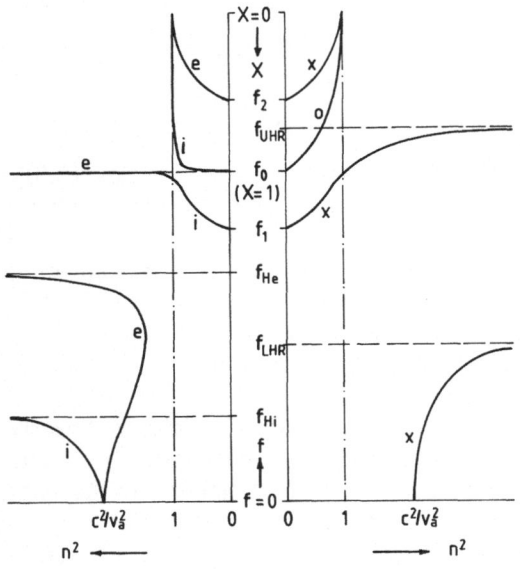

Quasi-Longitudinal   Quasi-Transversal

Propagation

Fig. 12.2. Dispersion of electromagnetic waves in cold plasma for quasi-longitudinal (*left*), and quasi-transversal (*right*) propagation. Plotted is the refractive index squared versus frequency f (or versus $X = f_0^2/f^2$). e electron waves; i ion waves; o ordinary waves; x extraordinary waves. For the meaning of the frequency symbols, see text

Waves with negative n^2 are attenuated along their propagation path according to $\exp(-|n|kz)$ [see Eq. (12.1)] and cannot transport wave energy. They are called evanescent waves and are of no importance.

The group velocity of the longitudinally propagating waves is

$$v_g = c/\{n + [2 \pm (Y_e - Y_i)](1 - n^2)^2/(2Xn)\}. \tag{12.9}$$

Of particular importance is the group velocity at frequencies approaching zero ($Y_e \gg Y_i \gg 1$; $X \gg 1$; $n \simeq c/v_a \gg 1$):

$$v_g \simeq v_a[1 \mp 1/(2Y_i)] \quad (X \to \infty), \tag{12.10}$$

and the group velocity of e-waves in the range $Y_e > 1 \gg Y_i$ [$n^2 \simeq X/(Y_e - 1)$]

$$v_g \simeq (2c/Y_e)[(Y_e - 1)^3/X]^{1/2}. \tag{12.11}$$

In the case of wave propagation transverse to an external magnetic field [transverse propagation; $B_0 = (0, B_0, 0)$], the refractive index becomes (Ratcliffe 1972)

$$n^2 = \begin{cases} 1 - X \\ [X - (1 + Y_e)(1 - Y_i)][X - (1 - Y_e)(1 + Y_i)]/[(1 - Y_e^2) \\ \quad \cdot (1 - Y_i^2) + X(Y_e Y_i - 1)]. \end{cases} \tag{12.12}$$

Figure 12.2 (right) shows the dispersion relation for transverse propagation. Here, we have the same three cutoff frequencies ($n = 0$) as in the case of longitudinal propagation [Eq. (12.7)], and two resonance frequencies ($n \to \infty$) at

$$X = (1 - Y_e^2)(1 - Y_i^2)/(1 - Y_e Y_i). \tag{12.13}$$

These frequencies are called upper (UHR) and lower (LHR) hybrid frequency. Because of $Y_i \ll Y_e$, they are given approximately by

$$f_{\text{UHR}} \simeq (f_0^2 + f_{He}^2)^{1/2}; \quad f_{\text{LHR}} \simeq [f_0^2 f_{He} f_{Hi}/(f_0^2 + f_{He}^2)]^{1/2}. \tag{12.14}$$

The phase velocity at low frequencies ($X \to \infty$) approaches v_a [see Eq. (12.8)].

If the ion pressure would have been taken into account, the phase velocity in the limit $X \to \infty$ for transverse propagation would have changed to $v_p = (v_a^2 + v_s^2)^{1/2}$, instead of Eq. (12.8), with v_s the velocity of sound (e.g., Denisse and Delcroix 1963). This wave is called the fast hydromagnetic wave. Introduction of thermal effects in the dispersion equation allows the possibility of electrostatic waves which can be excited near harmonics of the frequencies f_{He}, f_{Hi}, f_e, and f_i (e.g., Gendrin 1983).

Waves propagating transverse to the external magnetic field are linearly polarized, the field vector of one wave oscillating in the plane of the magnetic field, and the field vector of the other wave oscillating orthogonal to the magnetic field. At the hybrid frequencies, the electric field oscillates entirely in the direction of propagation (z-direction). The wave which has the same cutoff frequency as it would have in the absence of a magnetic field (f_0) is called the ordinary (o) wave. The second wave is called the extraordinary (x) wave.

If several ions are involved in the magnetospheric plasma (e.g., H^+, He^+ in addition to O^+), one must modify the dispersion equation. For instance, the refractive index for longitudinal propagation in a plasma with two ion components becomes, instead of Eq. (12.5),

$$n^2 = 1 - X_e/(1 \pm Y_e) - X_1/(1 \mp Y_1) - X_2/(1 \mp Y_2),\qquad(12.15)$$

where the indices "1" and "2" stand for the respective ion components. The dispersion equation now allows three solutions with crossing of two curves at a cross-over frequency f_c given by

$$1/(Y_c^2 - 1) = A_1/(Y_c^2 - M_1^2) + A_2/(Y_c^2 - M_2^2)\qquad(12.16)$$

with $Y_c = f_{He}/f_c$, and $A_j = n_j/n_e$; $M_j = m_j/m_e$ relative number density and relative mass of the j-th ion component.

Figure 12.3 shows this cross-over effect in the low frequency regime. The two wave types have the same phase and group velocities and the same polarization at the cross-over frequency f_c and can, therefore, exchange energy at this point.

Application of plane wave propagation, as discussed above, is possible if the propagation medium changes slowly along a distance of one wavelength of the wave. In this case, one needs to consider only the phase change along the propagation path, neglecting partial reflections of the wave. Instead of Eq. (12.1), the wave amplitude varies like

$$\exp\left[ik \int_h^z n(\theta)\cos(\theta - \xi)\,ds - i\omega t \right],\qquad(12.17)$$

where θ is the angle between magnetic field and wave normal, and ξ is the angle between magnetic field and ray direction (Fig. 12.4). This kind of approximation is called ray theory. For a wave packet propagating in the direction of ξ, it yields (e.g., Ratcliffe 1972)

$$\tan(\theta - \xi) = (1/n)\,\partial n/\partial \theta.\qquad(12.18)$$

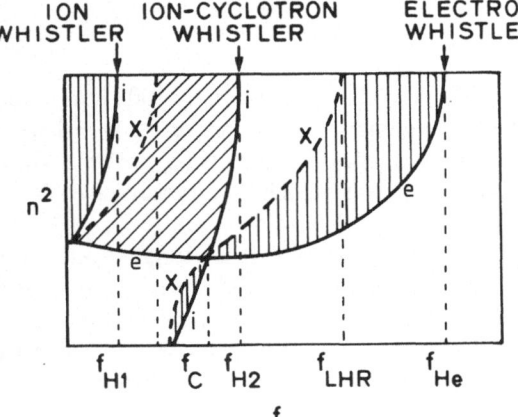

ION WHISTLER   **ION-CYCLOTRON WHISTLER**   **ELECTRON WHISTLER**

f_{H1} f_C f_{H2} f_{LHR} f_{He}

f

Fig. 12.3. Plot of n^2 versus frequency for a two-ion plasma. *Solid curves* are for quasi-longitudinal propagation, and *dashed curves* are for quasi-transversal propagation. f_{H1} and f_{H2} are the respective gyrofrequencies of the two ion components. f_c is the cross-over frequency, f_{LHR} is the lower hybrid frequency of component "1", and f_{He} is the electron gyrofrequency. (Park 1982)

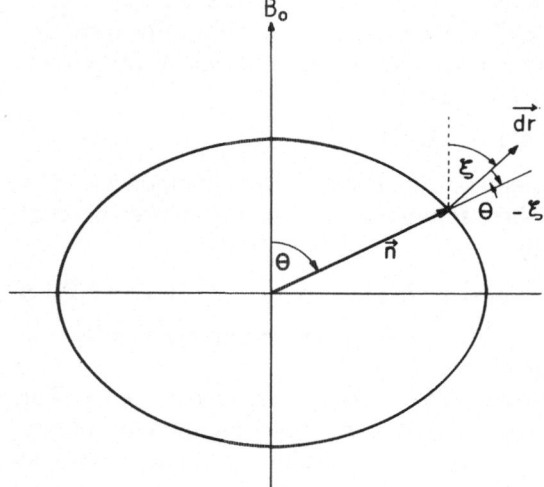

We have determined n for $\theta = 0$ in Eq. (12.5) and for $\theta = \pi/2$ in Eq. (12.12). For intermediate values of θ, it is a reasonable approximation to replace $Y_{e,i}$ by $Y_{e,i}\cos\theta$ in Eq. (12.5) (quasi-longitudinal propagation). For e-waves in the range $Y_i \ll 1 < Y_e$, with $n^2 \simeq X/Y_e$ in Eq. (12.5), one may write

$$\tan\xi = \tan\theta/(2 + \tan^2\theta) \leq (2)^{-3/2} \tag{12.19}$$

or $\xi_{\max} = 19.5°$. The ray direction of these waves can, therefore, differ by as much as about 20° from the magnetic field direction (e.g., Helliwell 1965).

12.2 Full Wave Theory

The magnetosphere behaves like a resonant cavity for waves with wavelengths comparable with the dimensions of the magnetosphere. The largest resonant wavelength in a cubic cavity of side length b (e.g., Magid 1972) is

$$\lambda_R = \sqrt{2}\,b, \tag{12.20}$$

which yields a period of $\tau_R \simeq \lambda_R/v_a \simeq 140\,\text{s}$ with $v_a \simeq 1000\,\text{km/s}$ and $b \simeq 15a = 10^5\,\text{km}$. The resonance frequency of this wave is $f_R = 1/\tau_R = 7\,\text{mHz}$, which lies well below the gyrofrequency of the ions ($f_{Hi} > 1\,\text{Hz}$ within the inner magnetosphere), so that the phase velocity of these waves is the Alfven velocity [see Eq. (12.8)].

Clearly, the ray approximation breaks down for these long wavelengths, and one has to take into account the real configuration of the magnetosphere. To date, no exact solution of this complicated problem exists. In the following, we try a highly simplified semi-analytical approach. We consider the electric and magnetic fields of a harmonic wave of angular frequency ω as small perturbations from a basic state and linearize the pertinent equations which are Maxwell's equations [neglecting the displacement current; Eqs. (14.11) and (14.12)]:

$$\nabla \times B = \mu j; \quad \nabla \times E = i\omega B \tag{12.21}$$

Ohm's law [assuming large electric conductivity; $\sigma \to \infty$; Eq. (14.20)]:

$$0 = E + v \times B_0 \tag{12.22}$$

and the momentum equation of the plasma (neglecting all mechanical forces):

$$-i\omega \varrho_i v = j \times B_0, \tag{12.23}$$

where B_0 is the geomagnetic field, assumed to be large compared with the magnetic field of the wave B.

We first consider pure poloidal oscillations of B [$B = (B_r, B_\theta, 0)$]. Poloidal oscillations are oscillations in the meridional plane of a spherical coordinate system. In this case, the plasma velocity v has also a poloidal component while the electric field and the electric current have only an azimuthal component [$E = (0, 0, E_\lambda); j = (0, 0, j_\lambda)$]. Since the geomagnetic field (approximated again by a coaxial dipole field) is also a poloidal field [see Eq. (14.3)], it follows from Eqs. (12.22) and (12.23) after vector multiplication with B_0

$$i\omega \varrho E = -B_0^2 j. \tag{12.24}$$

After elimination of B from Eqs. (12.21) and (12.24), we obtain

$$\nabla \times (\nabla \times E) = (\omega^2/v_a^2)E, \tag{12.25}$$

which, in the case of no azimuthal dependence ($\partial/\partial \lambda = 0$), can be rewritten (e.g., Dungey 1967) as

$$\nabla^2 E_\lambda + [\omega^2/v_a^2 - 1/(r^2 \sin^2 \theta)] E_\lambda = 0 \tag{12.26}$$

with ∇^2 the Laplace operator in spherical coordinates.

In order to find approximate solutions of this intractable equation, we assume that v_a is a constant, which is, of course, unrealistic (Fig. 12.5). Furthermore, we approximate the term $\sin^2 \theta$ in Eq. (12.26) by an average value of 0.5. The error due to this approximation will be large at polar latitudes, so that our approach will apply only to low and middle latitudes.

We solve Eq. (12.26) with these assumptions and find solutions in terms of separable variables:

$$E_\lambda = [A_n J_p(Kr/a) + B_n N_p(Kr/a)] (r)^{-1/2} P_n(\cos \theta) \exp(-i\omega t)$$

$$i\omega B_r = 1/(r \sin \theta) \partial/\partial \theta (\sin \theta E_\lambda) \tag{12.27}$$

$$i\omega B_\theta = -(1/r) \partial/\partial r(rE_\lambda),$$

with $P_n(\cos \theta)$ the Legendre polynomials, J_p, N_p Bessel and Neumann functions, A_n, B_n, integration constants, $K/a = \omega/v_a$ and $p = [(n + 1/2)^2 + 2]^{1/2}$, the factor of 2 in p arising from $1/\sin^2 \theta \simeq 2$.

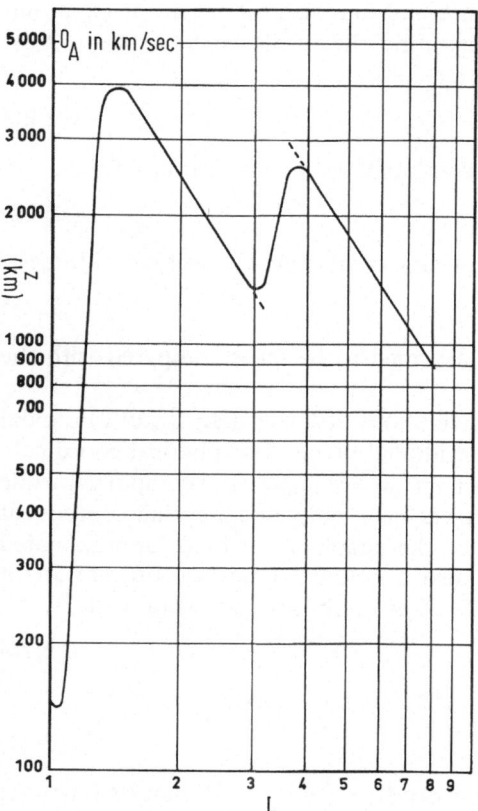

The integration constants are determined from the boundary conditions. The azimuthal electric field must disappear at the highly conducting ionospheric dynamo layer: $E_\lambda = 0$ at $r = r_D \simeq a$. The magnetic component orthogonal to the magnetopause must disappear. Since this boundary condition cannot adequately be met in our simplified approach, we assume that $B_r = 0$ at $r = b$. These boundary conditions lead to an eigenvalue equation

$$\begin{vmatrix} J_p(K) & N_p(K) \\ J_p(Kb/a) & N_p(Kb/a) \end{vmatrix} = 0 \tag{12.28}$$

and the ratio of the integration constants becomes

$$B_n/A_n = -J_p(K)/N_p(K) . \tag{12.29}$$

Anticipating the result that $K \ll 1$ and $Kb/a \gg 1$, one can use asymptotical formulas of the cylindrical functions (e.g., Menzel 1960) to obtain the following solution to the eigenvalue equation (12.28)

$$\tan[Kb/a - (p+1/2)\pi/2] \simeq -(K/2)^{-2p} \Pi(p)/[\Pi(-p)\sin p\pi] \tag{12.30}$$

with $\Pi(x)$ the factorial function. Since the right hand side of Eq. (12.30) is very large, a sufficiently accurate solution is

$$Kb/a - (p+1/2)\,\pi/2 \simeq (2m-1)\,\pi/2 \quad (m = 1, 2, \ldots)\,, \tag{12.31}$$

from which one obtains the eigenperiods

$$\tau_{nm} = 2\pi/\omega_{nm} \simeq 4b/[v_a(2m+p-0.5)]\,. \tag{12.32}$$

Using reasonable numbers in Eq. (12.32) ($n = m = 1$; $b \simeq 15a$; $v_a \simeq 1000$ km/s), one obtains $\tau_{11} \simeq 110$ s, a period which is consistent with the zero order estimate in Eq. (12.20). The wave components E_λ and B_θ of this mode have a latitude dependence given by $P_1(\cos\theta) = \cos\theta$ which is antisymmetric with respect to the equator. However, B_r is symmetric with respect to the equator. This kind of wave is called an even mode (Fig. 12.6b). A wave where B_r is antisymmetric is called an odd mode, e.g., the fundamental mode with $n = 0$ in Eq. (12.27). The magnetic field of the poloidal waves is purely meridional on the ground. The polarization of the even modes is equal on both hemispheres. It is opposite in the northern and the southern hemisphere for the odd modes.

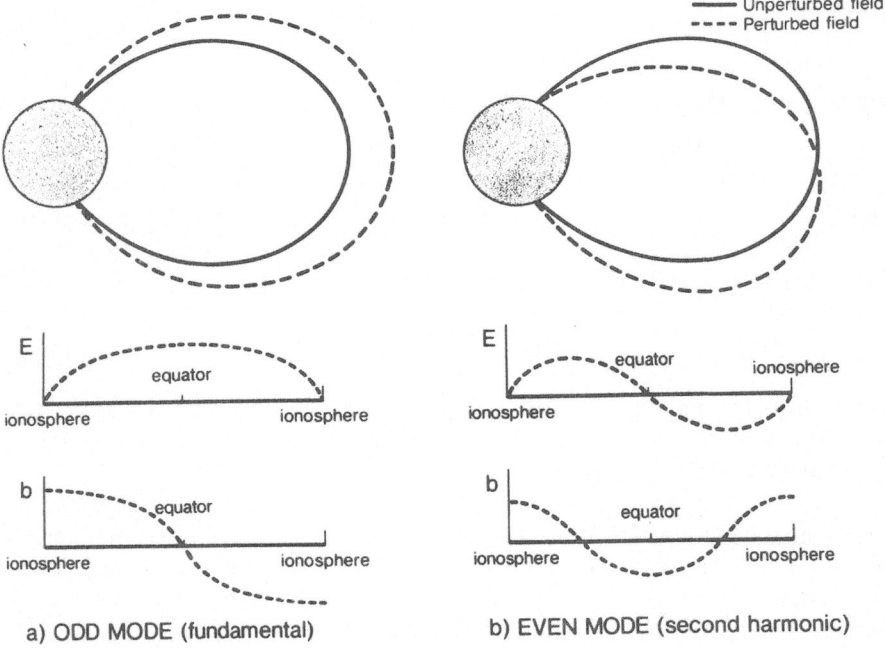

STANDING OSCILLATIONS IN A DIPOLE FIELD

Fig. 12.6. a Magnetic field line displacement (*top*), electric field amplitude along the background field (*middle*), and magnetic component (*bottom*) in the lowest odd standing mode (fundamental mode). **b** as for **a** for the lowest even standing mode. (Southwood 1981)

The second prototype of waves is a toroidal wave where the magnetic vector has only an azimuthal component $[B = (0, 0, B_\lambda)]$. The corresponding electric field has a poloidal component $[E = (E_r, E_\theta, 0)]$. This case is even more intractable than that of the poloidal wave. A WKB approximation is possible, however, which gives the resonance period in terms of the length of the (closed) geomagnetic field line (e.g., Rostoker 1979):

$$\tau_m \simeq (4/m) \int_{\theta_0}^{\pi/2} ds/v_a \quad (m = 1, 2, \dots), \tag{12.33}$$

with θ_0 the colatitude at which the field line intersects the ionosphere. This result can be interpreted as arising from a standing transverse oscillation of the field line as indicated in Fig. 12.6. The resonance period increases with increasing length of the field line, that is, with increasing latitude. Equation (12.33), certainly, breaks down in the range of the open field lines.

The observed geomagnetic pulsations (see Sect. 13.1) indeed show periods of the order of magnitude derived in this section, and their maximum amplitudes are near auroral latitudes. They are elliptically polarized, in general, indicating that poloidal and toroidal wave are coupled with each other in the real magnetosphere. Moreover, the waves are often located in narrow zonal bands.

12.3 Excitation of Plasma Waves

The main excitation mechanism of plasma waves within the magnetosphere is the interaction between highly energetic electrons and ions and the magnetospheric plasma. If a stream of energetic ionized particles with velocity v moves in the background plasma, it can excite plasma waves the phase velocity v_p of which equals the particle velocity:

$$v = v_p \quad \text{or} \quad \omega = kv, \tag{12.34}$$

where ω is the angular frequency of the waves and $k = \omega/v_p$ is their wave number. This type of wave excitation is called longitudinal resonance or Landau resonance.

A second type of waves can be generated if electrons (ions) gyrate along the geomagnetic lines of force with gyrofrequency ω_{He} (ω_{Hi}) and parallel velocity $v_{\parallel}$. The excited waves are circularly polarized e-waves (i-waves). Their frequency as seen by the moving particles must be Doppler-shifted to the gyrofrequency in order to achieve resonance:

$$\omega_H = \omega(1 - v_{\parallel}/v_p). \tag{12.35}$$

Since the frequency ω must be less than the gyrofrequency ($Y_e > 1$ for e-waves; $Y_i > 1$ for i-waves; see Fig. 12.2), the waves propagate in opposite direction to the particles, and the resonance condition becomes

$$(\omega_H - \omega)/\omega = |v_{||}|/v_p. \tag{12.36}$$

If the wave normal is not parallel to the magnetic field, higher-order resonances can occur according to the condiion

$$\omega = \pm v\omega_H + k_{||}v_{||} \quad (v = 0, 1, 2 \ldots), \tag{12.37}$$

where $v_{||}$ and $k_{||}$ are the respective components of v and k parallel to the magnetic field. Clearly, $v = 0$ corresponds to the longitudinal resonance [Eq. (12.34)], and $v = 1$ corresponds to the case in Eq. (12.36). Wave excitation for a wave number v, different from zero, is called transverse resonance or cyclotron resonance (e.g., Gendrin 1983).

Cyclotron resonance of e-waves is related to the parameters of the plasma as follows: from Eq. (12.36) together with the definition of the refractive index n approximated as in Eq. (12.11), one obtains the parallel kinetic energy of the energetic particles as

$$W_{||} = mv_{||}^2/2 = B_0^2/(2\mu n_e)(f_{He}/f)(1 - f/f_{He})^3, \tag{12.38}$$

where we have used the relationship $f_{He}^2/f_e^2 = B_0^2\varepsilon/(m_e n_e)$ from Eqs. (2.11) and (2.12) (Rycroft 1974).

In the magnetosphere, the flux of energetic particles tends to increase toward lower energies. The greatest contribution from the electrons to a given frequency along a given geomagnetic field line, therefore, is expected to come from the electrons of lowest energy that can satisfy Eq. (12.38). It is evident that, other things being equal, $f \approx f_{He}$. The term $B_0^2/(2\mu n_e)$ in Eq. (12.38) is smallest in the region where the field line crosses the equatorial plane. It is usually assumed, therefore, that the equatorial region is the main source of cyclotron resonance emission.

Waves already present within the magnetosphere can be amplified by energetic particles, mainly near the magnetospheric equator. On the other hand, waves may also trigger electron bursts which precipitate into the upper atmosphere (wave-particle interactions) (e.g., Thorne 1975; Scarf 1975; Shawhan 1979; Gendrin 1981; Southwood and Hughes 1983).

12.4 Ionospheric Screening Layer

In Section 12.2, the ionospheric dynamo region was considered to be infinitely well conducting so that waves of magnetospheric origin are totally reflected. In reality, the dynamo region is a moderately conducting layer, thin compared with the wavelength of the hydromagnetic waves. Some wave energy penetrates through this thin layer before it is partially reflected at the earth's surface.

We estimate here the reflection and transmission of the waves at this complex boundary consisting of the dynamo region, the middle and lower atmosphere, and the earth. The waves are Alfven waves in the magnetosphere propagating along the geomagnetic field lines which are assumed to be vertically directed. The dynamo region of thickness $\Delta h_D = h_M - h_D$ has a height-integrated Pedersen conductivity $\Sigma_p = \sigma_p \Delta h_D$. The Hall conductivity will be neglected for the moment. The middle and lower atmosphere reaching from 0 to about 100 km ($\Delta h_E = h_D - h_E$) is nonconducting. The earth has the constant conductivity σ_E. Vertical propagation of linearly polarized waves between heights h_1 and h_2 can be described by (e.g., Volland 1968)

$$c(h_2) = \exp[ikK(h_2 - h_1)]c(h_1) , \tag{12.39}$$

with $k = \omega/c$ the wave number in vacuo, $\omega \ll \sigma \mu c^2$, $c = \begin{pmatrix} E_x \\ B_y \end{pmatrix}$ the matrix of the wave amplitudes of the electric and magnetic field, and $K = \begin{pmatrix} 0 & 1 \\ n^2 & 0 \end{pmatrix}$ a coefficient matrix. n is the refractive index, which obtains the following forms in the various propagation regimes: $n = n_M \simeq c/v_a$ in the magnetosphere, $n = n_D \simeq (i\mu\sigma_p c^2/\omega)^{1/2}$ in the dynamo region, $n = 1$ in the middle and lower atmosphere, and $n = n_E \simeq (i\mu\sigma_E c^2/\omega)^{1/2}$ inside the earth. Since E_x and B_y must be continuous at a boundary, wave propagation from h_E (earth's surface) to h_M (lower boundary of the magnetosphere) is given by

$$c(h_M) = \exp[ikK_D(h_M - h_D) + ikK_E(h_D - h_E)]c(h_E) . \tag{12.40}$$

In order to determine reflection and transmission coefficients, one must transform from the field strength components to the characteristic upward- and downward-propagating waves a:

$$c = Qa , \tag{12.41}$$

with

$$a = \begin{pmatrix} a_0 \exp(iknz) \\ b_0 \exp(-iknz) \end{pmatrix}$$

and

$$Q = \begin{pmatrix} 1 & 1 \\ n & -n \end{pmatrix}$$

It can be shown (Volland 1968) that

$$S = \exp(ikKz) = \begin{pmatrix} \cos(knz) & i\sin(knz)/n \\ in\sin(knz) & \cos(knz) \end{pmatrix} . \tag{12.42}$$

Thus,

$$a_M = \begin{pmatrix} a_M \\ b_M \end{pmatrix} = Q_M^{-1} S_D S_E Q_E \begin{pmatrix} 0 \\ b_E \end{pmatrix} \tag{12.43}$$

relates the upgoing (a_M) and downgoing (b_M) magnetospheric waves with the wave penetrating into the earth (b_E).

If we use the parameters $\omega = 0.1 \text{ s}^{-1}$; $\Delta h_D = 50 \text{ km}$; $\Delta h_E = 100 \text{ km}$; $\Sigma_p = 10 \text{ S}$; $\sigma_E = 0.001 \text{ S/m}$; $v_a = 300 \text{ km/s}$; $\alpha = -in_D \sin(kn_D\Delta h_D) \simeq \mu\Sigma_p c = 4000$, then

$$S_D = \exp{(ikK_D\Delta h_D)} \simeq \begin{pmatrix} 1 & 0 \\ -\alpha & 1 \end{pmatrix}; \quad S_E = \exp{[ikK_E\Delta h_E]} \simeq \begin{pmatrix} 1 & 0 \\ 0 & 1 \end{pmatrix}$$

and

$$a_M = R\,b_M; \quad b_E = T\,b_m \tag{12.44}$$

with $R = (n_M - \alpha - n_E)/(n_M + \alpha + n_E) \simeq -0.87 - 0.11\,i$ a reflection factor, and $T = 2n_M/(n_M + \alpha + n_E) \simeq 0.03 - 0.03\,i$ a transmission factor. The lower and middle atmosphere have no influence on wave propagation of these long waves, and the ionospheric reflection and transmission factors are nearly independent of frequency within our approximation.

The reflection factor depends on the refractive indices of earth, ionosphere and magnetosphere. If their magnitudes are nearly equal ($n_M \simeq \alpha \simeq |n_E|$), R may decrease to 0.45 and T may increase to 0.90. This may happen for $\omega = 0.5\,\mathrm{s}^{-1}$; $\sigma_E = 10^{-4}\,\mathrm{S/m}$; and $v_a = 100\,\mathrm{km/s}$, numbers well within the range of observed values (see also Greifinger and Greifinger 1965).

We now estimate the influence of the anisotropy of the dynamo region on wave propagation. In the simplest case of longitudinal propagation, the wave polarization is subject to Faraday rotation. Faraday rotation can be understood as arising from the different phase velocities of the circularly polarized i-waves and e-waves. A linearly polarized wave from the magnetosphere penetrating the anisotropic dynamo region splits into an e-wave and an i-wave with refractive indices

$$\left. \begin{array}{c} n_i \\ n_e \end{array} \right\} \simeq n_D(1 \mp i\beta)^{1/2}, \tag{12.45}$$

with $n_D = (i\mu\sigma_p c^2/\omega)^{1/2}$; $\beta = \sigma_h/\sigma_p$; and σ_h the Hall conductivity. After penetrating through the dynamo region, both waves superpose again, resulting in a linearly polarized wave which is shifted in phase by

$$\phi = [\omega/(2c)]\Delta h_D \operatorname{Real}(n_i - n_e). \tag{12.46}$$

Using the numbers from above and $\beta = 1$, one obtains $\phi \simeq 5°$. This phase shift increases with frequency and with ionospheric conductivity. More sophisticated calculations taking into account the limited longitudinal spacing of the geomagnetic pulsations show that the phase shift may be as large 90° (Hughes and Southwood 1976). This has been confirmed by observations (e.g., Andrews et al. 1981).

e-Waves in the VLF range can penetrate the magnetosphere and propagate along the geomagnetic field lines with group velocity according to Eq. (12.11). These circularly polarized waves with CW polarization suffer attenuation in the lower dynamo region due to electron-neutral collisions. If one includes collisions, the refractive index for longitudinal propagation of e-waves changes from Eq. (12.5) to ($Y_e > 1 \gg Y_i$)

$$n_e \simeq [1 + X_e/(Y_e - 1 - iZ_e)]^{1/2} \simeq (X_e/Y_e)^{1/2}[1 + iZ_e/(2Y_e)], \tag{12.47}$$

where $Z_e = \nu_e/\omega$, and ν_e is the electron-neutral collision frequency. The attenuation rate (in decibels) is given by

$$\gamma = 8.7\, Im[kn_e\Delta h_D] \simeq 8.7 f_e \nu_e \Delta h_D (f/f_{He})^{1/2}/(2f_{He}c)\,. \tag{12.48}$$

The attenuation is strongest in the regions where $Z_e > 1$, which is within the D layer and lower E layer below about 100 km altitude.

Typical daytime values are 6 dB for $f = 1.5$ kHz, decreasing to about 1.3 dB during nighttime conditions (Helliwell 1965). γ increases with frequency and also with increasing angle of incidence of the waves.

12.5 Transmission Line Model of Magnetosphere

The electric current system during a substorm, outlined in Section 11.7, together with the existence of pulse type disturbances (Fig. 11.12) suggests an equivalent transmission line for the electrojet circuit in Fig. 11.13 of the form of Fig. 6.5. A voltage U (corresponding to the fluctuating electric convection potential) drives electric currents during substorms, which are closed via field-aligned currents and the ionospheric dynamo region. We attribute a total resistance $R' = R_f/d$ to the geomagnetic field lines, where d is the length of the diverted field-aligned current sheets in the tail. The conductance G', which is a measure of the conductivity orthogonal to the magnetic field lines, will be neglected. The mutual electric coupling of the field-aligned currents is simulated by an inductance $L' = L_f/d$ [Eq. (11.45)], and the energy storage in the near tail plasma is simulated by a capacitance $C' = C_f/d$ [Eq. (11.47)]. The lower input terminal at $z = 0$ (at the dynamo region) is bridged by the input conductance $G_e = 1/R_e$, where R_e is the resistance of the nighttime aurora (Fig. 11.13). The relationship between voltage and current (Magid 1972) is (for convenience, we drop the suffices "f" at R_f, L_f, and C_f)

$$d\,\partial U/\partial z = RJ + L\,\partial J/\partial t; \quad d\,\partial J/\partial z = C\,\partial U/\partial t\,, \tag{12.49}$$

with the lower boundary condition

$$J(0, t) = G_e U(0, t)\,. \tag{12.50}$$

We ask for a solution of the following problem: How does this wave guide react to a sudden voltage pulse at its upper end? We select as initial condition at $z = d$ (the magnetopause) a delta function $\delta(t)$:

$$U(d, t) = U_0\delta(t)\,, \tag{12.51}$$

with U_0 a constant amplitude. After a Fourier transformation of U, one obtains

$$U(z, t) = 1/(2\pi) \int_{-\infty}^{\infty} d\omega \int_{-\infty}^{\infty} [A(\xi)\exp(iKz/d)$$

$$+ B(\xi)\exp(-iKz/d)]\exp[i\omega(\xi - t)]\,d\xi\,, \tag{12.52}$$

with K from Eq. (6.5) ($G' = 0$). $A(\xi)$ and $B(\xi)$ are determined from the initial condition (12.51) at $z = d$ and from the boundary condition (12.50) at $z = 0$. The final solution of U at the lower boundary ($z = 0$) is

$$U(0, t) = U_0/(2\pi) \int_{-\infty}^{\infty} KG_e \exp(-i\omega t) d\omega/(KG_e \cos K + i\omega C \sin K)$$

$$\simeq i U_0 \Sigma c_\nu \exp(-i\omega_\nu t), \tag{12.53}$$

where c_ν are the residues in the Laurent development of the integral.

The eigenvalues ω_ν are found from the zeros in the denominator of the integral (12.53) as

$$i\omega_\nu = \gamma_\nu \pm i\delta_\nu, \tag{12.54}$$

with $\gamma_\nu \simeq R/(2L)$; $\delta_\nu \simeq (K_\nu^2 \omega_0^2 - \gamma_\nu^2)^{1/2}$; $K_\nu \simeq (\nu + 1/2)\pi$ ($\nu = 1, 2, 3, \ldots$).

$$\omega_0 = 1/(LC)^{1/2} = (\gamma_\nu^2 + \delta_\nu^2)^{1/2}/K_\nu \tag{12.55}$$

is the eigenfrequency of the system. The phase velocity of the wave modes is

$$v_p = \delta_\nu d/K_\nu. \tag{12.56}$$

13 Waves in the Magnetosphere

In this chapter, we outline low frequency wave phenomena within the magnetosphere. These waves are generated within the magnetosphere, or they are excited by lightning events and propagate along the geomagnetic lines of force as whistlers.

13.1 Geomagnetic Pulsations

Fluctuations of the geomagnetic field with periods ranging from 0.2 s to more than 10 min are called geomagnetic pulsations. Table 13.1 gives a numerical classification of the observed pulsations. One differentiates between quasi-periodic pulsations (pc; p = pulsation; c = continuous) and irregular pulsations (pi; i = irregular). A physical classification according to their physical characteristics divides the numerical classes. We will give in this section a brief summary of the observations and their interpretations. For more details, see Saito (1978) and Rostoker (1979).

Figure 13.1 shows typical pc-events. Figure 13.2 gives an example of a $pi2$-event associated with the onset of a substorm — a damped wave train lasting for a few cycles (Rostoker 1979; Heacock and Hunsucker 1981).

$Pc1$ pulsations of the pearl type (pp) having periods between 0.2 and 5 s are observed at all geographic latitudes. Their maximum occurrence is at sub-auroral latitudes. Their amplitudes range from 0.05 to 0.1 nT. In general, they occur after a geomagnetic storm. Their wave forms are those of a wave packet with rising tones. Their polarization is similar to poloidal waves with odd wave numbers (even modes), and the major axis of their polarization ellipses is preferentially orientated along the north—south direction.

A frequency—time diagram is shown in Fig. 13.3. The behavior in Fig. 13.3 can be interpreted as arising from wave pulses travelling back and forth along geomagnetic field lines, reflected at the ionospheric dynamo region. The group velocity of a pulse decreases with frequency [see Eq. (12.10); upper sign] so that the high frequency wing arrives later than the low frequency wing of the pulse. They are excited by cyclotron resonance within the magnetospheric equator regions as discussed in Section 12.3.

$Pc2$, $pc3$, and $pc4$ events (5 – 150 s) are mainly a daytime phenomenon. They tend to occur more often in the morning sector than in the afternoon

Table 13.1. Classification of magnetic pulsations. (Saito 1978)

Numerical Classification			Physical Classification			
Wave form	Period range (s)	Type	Type	Name	L^a	D^b
c o n t i n u o u s	0.2 – 5	$pc1$	pp	Pearl pulsation	S	
			hmc	Hydromagnetic chorus	A	D
			ce	Continuous emission	A	
			$ipdp$	Interval of pulsations, diminishing period	S	E
				Others		
	5 – 10	$pc2$	aip	Auroral irregular pulsation	A	M
				Others		
	10 – 45	$pc3$	$pc3$	$pc3$	A	M – D
				Others		
	45 – 150	$pc4$	$pc4$	$pc4$	S	D
			pg	Giant pulsation	A	M
				Others		
	150 – 600	$pc5$	$pc5$	$pc5$	A	D
				Others		
	>600	$pc6$	tf	Tail fluttering	T	N
				Others		
I r r e g u l a r	1 – 40	$pi1$	spt	Short-period pulsation train	A	N
			pib	pi Burst	A	N
			pic	pi (Continuous)	A	M
			pid	Daytime pi	A	D
			$psc1,2,3$	sc-Associated $pc1,2,3$	A	D
			$psi1,2,3$	si-Associated $pc1,2,3$	A	D
				Others		
	40 – 150	$pi2$	$pi2$	$pi2$ (Formerly pt)	A	N
			$psfe$	sfe-Associated pulsation	L	D
			$psc4$	sc-Associated $pc4$	S	D
			$psi4$	si-Associated $pc4$	S	D
				Others		
	>150	$pi3$	$psc5,6$	sc-Associated $pc5,6$	A	D
			$psi5,6$	si-Associated $pc5,6$	A	D
			pip	Polar irregular pulsation	A	N
			$ps6$	Substorm associated long-period pulsation	A	N
				Others		

[a] Latitude dependence of maximum amplitude (A: auroral zone, S: subauroral zone; L: low latitudes; T: magnetotail).
[b] Diurnal dependence of maximum amplitude (M: morning sector; D: daytime sector; E: evening sector; N: nighttime sector).

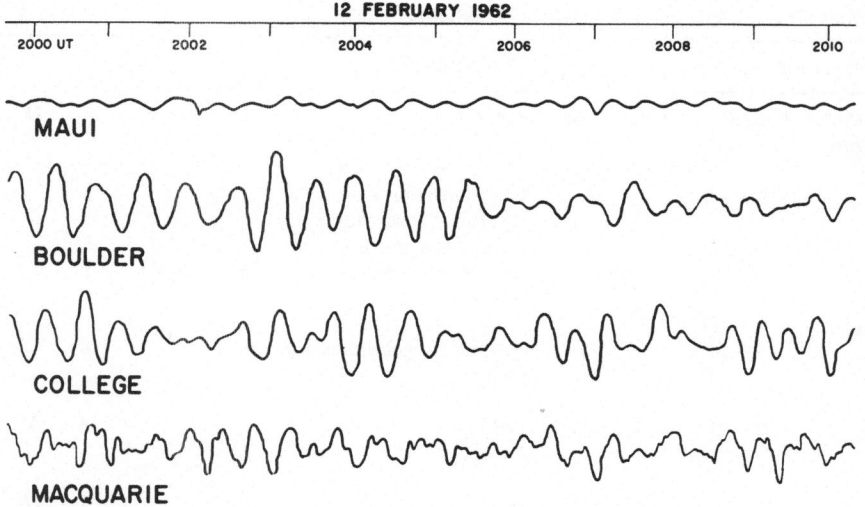

Fig. 13.1. Examples of geomagnetic $pc2$ pulsations showing hemispherical similarities. Macquarie Island and College, Alaska, are near conjugate stations in the auroral regions. Boulder, Colorado, is at midlatitudes, and Maui, Hawaii, is subequatorial. (Campbell 1967)

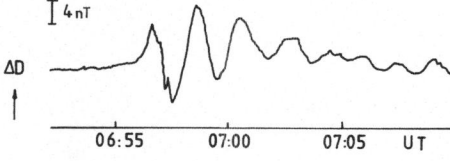

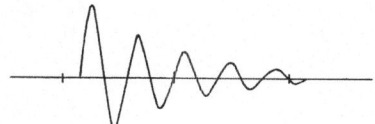

Fig. 13.2. Typical $pi2$ pulsation during substorms (*upper panel*) (Rostoker 1979); and theoretical curve (*lower panel*)

sector, with maximum amplitudes in the auroral ovals and near the plasmapause. A single event is a conjugate phenomenon, appearing synchronously at the opposite ends of a geomagnetic field line. There is a clear dependence of the average period on geomagnetic activity at midlatitude stations. During quiet conditions, ($Kp < 2$), $pc4$-events dominate. During disturbed conditions ($Kp > 4$), $pc2$-events occur more frequently.

$Pc5$-events (150 – 600 s) have large amplitudes (up to several hundreds of nanoteslas) and can persist for 1 h or more. They are most frequent in the prenoon hours. Their center of activity follows closely the auroral oval and lies within the boundaries of the westward polar electrojet. Their greatest intensity is during the recovery phase of magnetospheric substorms. $Pc5$ events in the afternoon sector are more of the form of damped pulsation trains ($pi3$).

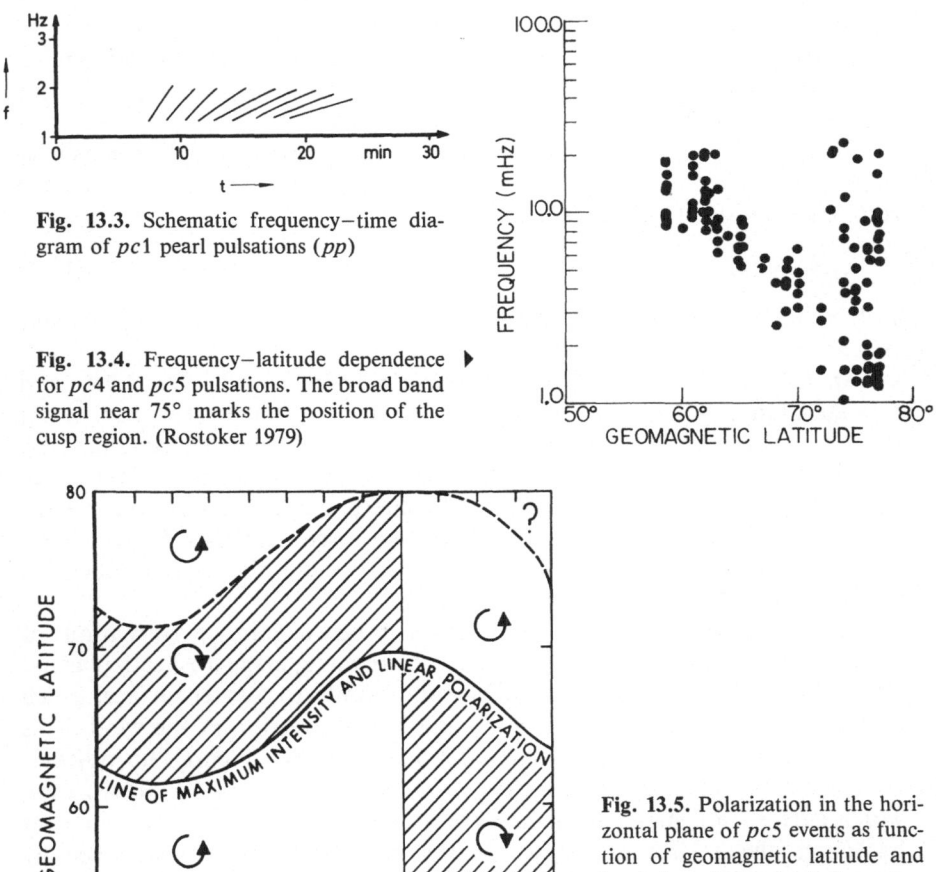

Fig. 13.3. Schematic frequency–time diagram of $pc1$ pearl pulsations (pp)

Fig. 13.4. Frequency–latitude dependence ▶ for $pc4$ and $pc5$ pulsations. The broad band signal near 75° marks the position of the cusp region. (Rostoker 1979)

Fig. 13.5. Polarization in the horizontal plane of $pc5$ events as function of geomagnetic latitude and local time. This plot is for pulsations with periods near 200 s. For shorter periods, the entire pattern shifts equatorward, and for longer periods, it shifts poleward. (Rostoker 1979)

Single $pc5$ oscillations are sometimes azimuthally located and decrease in intensity from maximum amplitudes to near noise level within one time zone (<650 km at auroral latitudes).

Simultaneous observations show a frequency–latitude dependence of $pc4$ and $pc5$ events as shown in Fig. 13.4 with a tendency for longer periods to occur at higher latitudes.

The sense of polarization of $pc5$-events is indicated in Fig. 13.5. For north-hemispheric midlatitude stations equatorward of the auroral oval, the polarization is CCW in the prenoon hours and CW in the afternoon hours. Within the auroral oval, the polarization is CW in the prenoon hours and CCW in the afternoon hours. In the polar cap, it is CCW. The polarization changes sign at conjugate points in the southern hemisphere. For $pc3$ and $pc4$ events, the entire pattern of Fig. 13.5 is shifted equatorward, so that the line of maximum intensity and of the reversal in the sense of polarization is near the plasmapause (50° to 60° latitude) for $pc3$.

$Pc2$ to $pc5$ events can be interpreted as arising from toroidal eigen oscillations on localized L shell regimes in the magnetosphere. We estimate the resonance periods from Eq. (12.33) as

$$\tau_m \simeq 2S/(m\bar{v}_a) \quad (m = 1, 2, 3, \ldots), \tag{13.1}$$

with S the length of the field line [e.g., from Eq. (14.10)] and $\bar{v}_a$ the average Alfven velocity along the field line.

At 60° latitude ($\theta \simeq 30°$), we use $L \simeq 1/\sin^2\theta = 4.0$, $S \simeq 9.0\,a$, and $\bar{v}_a \simeq 2000$ km/s to obtain $\tau \simeq 60$ s (or $f = 1/\tau = 17$ mHz) for $m = 1$. At $\theta = 20°$, however, it is $S \simeq 21.6\,a$ and $\tau \simeq 140$ s ($f \simeq 7$ mHz). This qualitatively explains the frequency dependence in Fig. 13.4. The broadening of the range of f in Fig. 13.4 near 75° latitude may be interpreted as arising from plasma instabilities at the boundary between open and closed field lines. The broadening of f near 60° latitude may be due to the increase of the Alfven velocity in the environment of the plasmapause (see Fig. 12.5).

The $pc2$ to $pc5$ oscillations are believed to be excited by perturbations of field lines at the magnetopause by the downstreaming magnetosheath plasma. The situation is analogous to a wind blowing parallel to a water surface, which results in the generation of waves moving in the direction of the wind. This mechanism is known as the Kelvin-Helmholtz instability. Figure 13.6 indicates this excitation mechanism, together with the expected wave polarization which agrees, in general, with the observations. Other possible mechanisms based on plasma instabilities within the magnetosphere are discussed by Rostoker (1979).

$Pc6$-events with periods larger than 600 s can persist for more than an hour. They are observed in the tail regions of the magnetosphere and are probably associated with a fluttering of the magnetic tail fields.

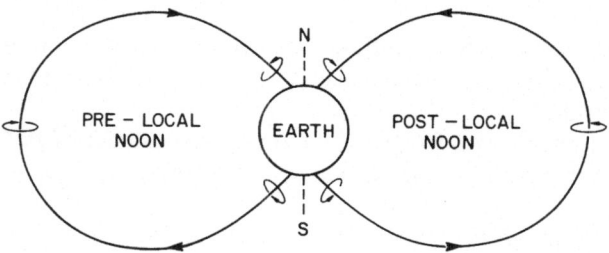

Fig. 13.6. Rotation of the polarization vector (as seen from the sun) of the transverse hydromagnetic mode as would be expected for the Kelvin-Helmholtz excitation mechanism on the day side of the magnetosphere. Counterclockwise polarization is expected during pre-noon, Clockwise polarization is expected during post-noon. (Lanzerotti and Southwood 1979)

Substorm associated $pi2$-events (pt) are primarily a night time phenomenon with maximum activity in the auroral ovals. At midlatitudes, the sense of polarization in the horizontal plane is CCW in the northern hemisphere and CW in the southern hemisphere. Their frequency-latitude dependence is similar to that in Fig. 13.4.

This type of $pi2$ can be explained as resonant oscillation of the magnetospheric transmission line, discussed in Section 12.5. For instance, the pulsation train in Fig. 13.2 (upper panel) can be interpreted as the first mode ($\nu = 1$) of the type 2 wave form in Eq. (6.16) with the parameters $\gamma_1 \simeq 4.4 \times 10^{-3}$ s^{-1}; $\delta_1 \simeq 5.2 \times 10^{-2}$ s^{-1} from Eq. (12.54). The lower panel in Fig. 13.2 is calculated with these parameters. The eigenfrequency of the magnetosphere according to Eq. (12.55) is then $\omega_0 \simeq (\gamma_1^2 + \delta_1^2)^{1/2}/K_1 = 0.011$ s^{-1}. This number is quite consistent with the corresponding value $\omega_0 = 1/(LC)^{1/2} \simeq 0.01$ s^{-1}, derived from the parameters $L_f = 30$ H; $C_f = 300$ F, as estimated in Section 11.7. The phase velocity of this mode is $v_p = \delta_1 d/K_1 \simeq 1000$ km/s (with $d = 15\,a$), consistent with the Alfven velocity in this region. However, the resistance, which is $R = 2L\gamma_1 \simeq 0.26$ Ω, is a factor of about 10 larger than the value R assumed in the substorm model in Section 11.7. This suggests that $pi2$ during substorms are excited when anormally high resistivity of the auroral field lines exists. This may occur mainly during a sudden change of state of the magnetospheric tail region. Anormal resistivity and parallel electric fields have, in fact, been observed along auroral field lines (Mozer 1981). For a review of $pi2$, see Baumjohann and Glassmeier (1984).

$Pi1$-events (spt) are observed at low and high latitudes and tend to occur with the onset of substorms. They appear as high frequency riders on the associated $pi2$ pulsations. In the language of transmission line theory, they can be considered as the higher order modes.

Another type of $pi2$ is related to rapid changes in the ionospheric conductivity during solar flare events with enhanced XUV fluxes ($psfe$). This type of $pi2$ is not associated with solar wind or magnetospheric disturbances (e.g., Rosenberg et al. 1981).

Sudden pulses (si) and sudden commencements (sc) are worldwide phenomena occurring nearly simultaneously on the day side and on the night side. They are believed to be excited by a sudden impact of the solar wind stream on the magnetopause (e.g., Nishida 1978).

Substorm-associated $pi3$, called $ps6$, are observed in the region of the westward polar electrojet during substorms. They are accompanied by auroral pulsations and by a wavy pattern of the electrojet. This pattern drifts from the midnight sector toward dawn and dusk during the expansion phase of the substorm (Saito 1978; Nagano et al. 1981).

13.2 Whistlers

The low frequency spectrum of the electromagnetic energy of lightning signals is ducted within the terrestrial waveguide (see Sect. 7.3). Part of this energy can tunnel through the ionosphere and can propagate in the e-wave mode through the magnetosphere. The allowed frequency band for this magneto-spheric propagation is limited by the electron gyrofrequency f_{He} or the electron plasma frequency f_e, whichever is larger for longitudinal propagation, and by the lower hybrid frequency f_{LHR} for transverse propagation (Fig. 12.2). Both f_{He} and f_{LHR} decrease with increasing L shell and reach values of about 50 kHz and 200 Hz, respectively, near the equatorial plasmapause.

The broad spectrum of a lightning pulse with maximum spectral amplitudes near 5 to 10 kHz (Fig. 7.10) is dispersed during its propagation through the magnetospheric plasma. In the case of longitudinal propagation, the group travel time along a geomagnetic line of force is

$$\tau = 2 \int_{\theta_0}^{\pi/2} ds/v_{gr} \simeq 1/(c\sqrt{f}) \int_{\theta_0}^{\pi/2} f_e ds/[\sqrt{f_{He}}\,(1 - f/f_{He})^{3/2}] \qquad (13.2)$$

with v_{gr} defined by Eq. (12.11).

The ground-based observation of the frequency–time signature of whistlers is shown schematically in Fig. 13.7. The group time τ decreases with increasing frequency below the "nose" frequency f_n. Above f_n, τ increases with frequency and asymptotically approaches infinity for $f \rightarrow f_{He}$. The nose frequency can be determined from $\partial\tau/\partial f = 0$ approximately as $f_n \simeq f_{He}/4$. Some representative values for f_n are 14 kHz at $L = 3$ and 1.5 kHz at $L = 6$.

The group time depends on the electron number density n_e along the propagation path because $f_e = 8.98\sqrt{n_e}$ (f_e in Hz; n_e in m^{-3}). Whistler measurements therefore enable a determination of the electron content along an L shell. The plasmapause, in particular, with its abrupt decrease in electron density at the last closed electric equipotential shell (Fig. 11.6) has been detected from whistler observations (e.g., Carpenter and Park 1973).

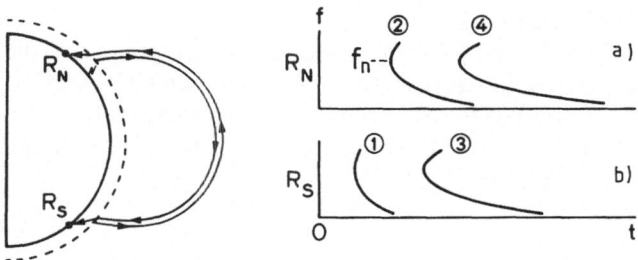

Fig. 13.7a, b. Whistler path along a geomagnetic field line (*left*), and frequency–time traces of the whistler (*right*). R_S is a one-hop whistler, R_N is a two-hop whistler. f_n is the nose frequency. (Park 1982)

Only the frequency range below the nose frequency is seen during most whistler observations. The frequency range near and beyond the nose frequency is normally strongly absorbed within the ionospheric D layer below about 90 km [Eq. (12.48)]. Moreover, the spectral amplitude of the lightning signal decreases with frequency above about 10 kHz (see Fig. 7.11).

For the low frequency range, Eq. (13.2) can be simplified to

$$D = \tau \sqrt{f} \simeq 8.94 \times 10^{-14} \overline{(n_e/B_0)^{1/2}} S \tag{13.3}$$

with $f_{He} = 2.80 \times 10^{10} |B_0|$ (f_{He} in Hz; B_0 in T). S (in m) is the length of the field [e.g., from Eq. (14.10)]. D (in $\sqrt{s}$) is a measure of the dispersion which is independent of frequency. A typical value for midlatitudes is $D \simeq 70$ corresponding to an average electron number density of $n_e \simeq 5 \times 10^9 \, \mathrm{m}^{-3}$, an average geomagnetic field of $B_0 \simeq 4 \, \mu\mathrm{T}$, and a path length of $S \simeq 2.2 \times 10^4 \, \mathrm{km}$ ($L = 2$).

Whistlers can be reflected at conjugate points and can, therefore, echo from hemisphere to hemisphere. The dispersion effect is enhanced with increasing path length. Whistlers originating near the observing station and their echoes have a frequency–time diagram as shown in Fig. 13.7a (two-hop whistler). Whistlers originating at the conjugate point have a frequency-time diagram as shown in Fig. 13.7b (one-hop whistler). Up to ten echos and more have been observed, on occasion.

The effect of the ionosphere is to damp the higher frequencies more strongly during the daytime than during the night [Eq. (12.48)]. Part of the whistler energy may be guided in the terrestrial wave guide between earth and ionosphere so that the whistler signal observed at a station does not necessarily come along that field line which has its foot print at the station.

It was shown in Section 12.1 [Eq. (12.19)] that the ray direction can differ as much as 20° from the geomagnetic field direction. The rays cannot be closely guided along the geomagnetic field lines. Although spatial gradients of the refractive index tend to rotate the wave normal toward the geomagnetic field, this adjustment may not be rapid enough to follow the change in field line direction. Under certain conditions, the wave normal may become orthogonal to the magnetic field line. If the wave frequency equals the lower hybrid frequency at this point, reflection as indicated in Fig. 13.8a is possible. Whistlers that undergo such reflections are called magnetospherically reflected (MR) whistlers. In situ receivers on satellites can measure whistlers which follow these complicated ray paths (e.g., Edgar 1976).

Whistlers observed on the ground such as in Fig. 13.7 are probably ducted along field-aligned plasma density irregularities. Figure 13.8b schematically indicates this ducting process. The life time of the ducts may vary from a few minutes to a few hours. How these ducts are formed is not well understood.

Whistler ducts have leaky surfaces allowing waves to be ducted along only part of their length. Unducted and partially ducted whistlers are strongly affected by the presence of ions, and they can propagate in a variety of modes depending on wave frequency, electron density, and ion composition. These whistlers can only be observed in space.

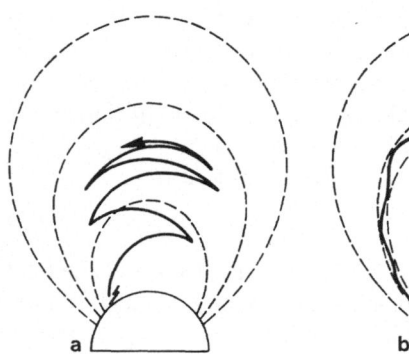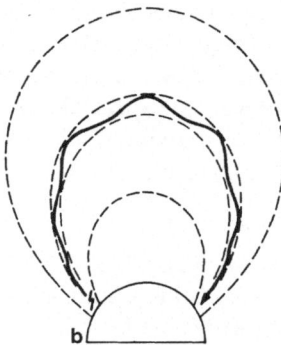

Fig. 13.8a, b. Illustration of different ray paths for ducted (*right*), and unducted (*left*) whistler-mode propagation. (Park 1982)

The magnetospherically reflected (MR) whistlers mentioned above are one type of whistler influenced by ions. Another type is the ion-cyclotron whistler phenomenon, which can be generated in regions where at least two ion constituents are present in sufficient number density. Its dispersion curve is the curve labeled "ion cyclotron whistler" in Fig. 12.3. This whistler mode can be excited by a normal whistler (sometimes called an electron whistler) if its frequency reaches the cross-over frequency in Fig. 12.3. At this point, part of the electron whistler energy is coupled into the ion-cyclotron mode. If the plasma consists of O^+ and H^+ ions, the ion-cyclotron mode is called a proton-cyclotron whistler. If the ion He^+ is also present, a second mode, called a helium-cyclotron whistler, exists. Ion-cyclotron whistlers cannot propagate to the ground because the H^+ and He^+ densities are too small in the lower part of the ionosphere. The ion whistler mode in Fig. 12.3 is due to the main plasma constituent in the ionosphere, O^+. Pearl (*pp*) pulsations excited within the magnetosphere, as discussed in Section 13.1 (Fig. 13.3), also obey this dispersion curve.

Figure 13.9b schematically shows the observed frequency–time behavior of an ion-cyclotron whistler measured at satellite heights. Figure 13.9a gives an explanation of this phenomenon. The curve on the right is the ion gyrofrequency of H^+ or He^+ as a function of altitude. The frequency indicates the upper limit of the allowed range of the ion-cyclotron whistler (see Fig. 12.3). The curve on the left shows the cross-over frequency vs height where energy coupling from the electron whistler mode into the ion-cyclotron mode is possible. The vertical thin lines represent spectral electron whistler waves propagating upward and delayed in time due to the dispersive effect of the magnetosphere. When these spectral frequencies coincide with the cross-over frequency, they couple part of their energy into the ion-cyclotron mode. The ion-cyclotron waves, thus excited, propagate upward, as indicated by the heavy arrows. However, their phase velocity is much slower than that of the electron whistlers because their refraction index is larger (Fig. 12.3). The ion-cyclotron whistler waves cannot exist beyond the ion gyrofrequency. A satellite moving along the horizontal dashed line, therefore, receives signals as indicated in Fig. 13.9b.

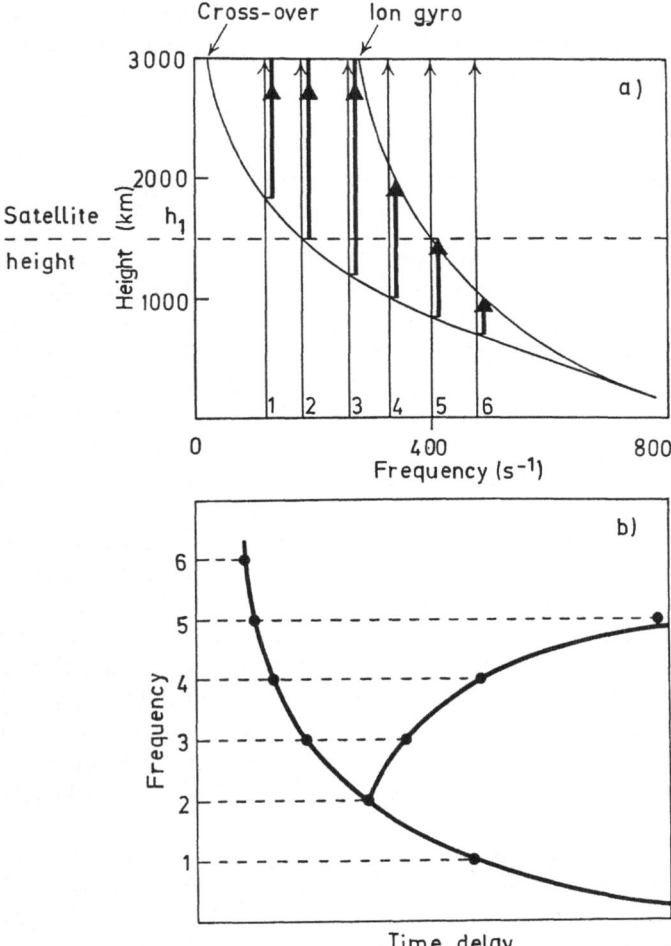

Fig. 13.9 a, b. Generation mechanism of ion-cyclotron whistler. **a** variation of the second ion gyrofrequency f_{H2} and of the cross-over frequency f_c [see Eq. (12.16) and Fig. 12.3]. *Thin vertical lines* represent waves on frequencies 1 to 6 produced by a lightning discharge and travelling upward in the electron-whistler mode. They excite additional waves (*thick vertical lines*) in the ion-cyclotron mode at the cross-over levels. The ion-cyclotron waves (*thick lines*) travel more slowly than the electron-whistler waves (*thin-lines*) and cannot penetrate above the level where their frequency is the same as f_{H2}. **b** Frequency–time trace that would be recorded in a satellite at the height shown by the *dashed line* in **a**. (Ratcliffe 1972)

Ion-cyclotron whistlers provide informations about the local ion composition (e.g., Gurnett et al. 1965). Other types of whistler phenomena at satellite heights are discussed by Park (1982).

13.3 Electromagnetic Waves of Magnetospheric Origin

In this section, we consider low frequency electromagnetic waves ($f \lesssim 300$ kHz) which are excited within the magnetosphere. Naturally occurring VLF signals of magnetospheric origin observed on the ground have a wide variety of tonal characteristics, but are distinctly different from whistlers. Figure 13.10 schematically shows some typical frequency–time signatures of these signals. Rocket and satellite measurement reveal a number of other plasma wave types. Figure 13.11 shows an overview of the most probable locations of the sources of plasma wave occurrence within the magnetosphere. Figure 13.11 includes the hydromagnetic waves discussed in Section 13.1.

Most of these wave types propagate in the whistler mode along the geomagnetic field lines, so that their wave characteristics are modified by dispersion. However, some waves are trapped (e.g., evanescent electrostatic waves with group velocity near zero). Others can propagate across the geomagnetic field (e.g., MR whistler modes as discussed in Sect. 13.2), or they can escape from the magnetosphere into space. Indeed, the magnetospheric regions aligned to the auroral ovals are copious sources of radio emission, which radiate low frequency wave power of about 10^9 W into space at frequencies greater than f_{He}, or f_e. Their maximum spectral amplitudes are near 200 kHz (Gurnett 1976; Barbosa 1982).

The noise observed on the ground can be divided into two broad categories. One category, called "chorus", shows clearly discernable discrete tones (Fig. 13.10). The second category, called "hiss", sounds like amorphous broadband noise. Many subcategories can be identified, and some cases are difficult to classify at all (e.g., Park 1982).

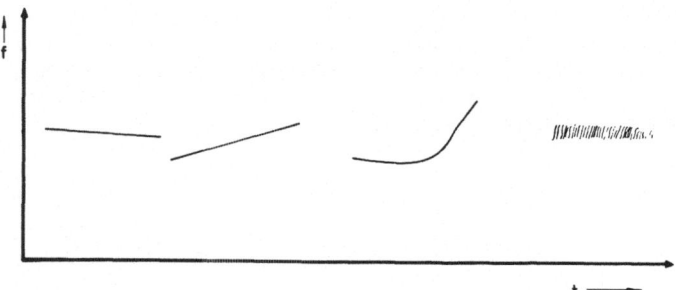

Fig. 13.10. Typical frequency–time signatures of discret tones of magnetospheric origin (schematic)

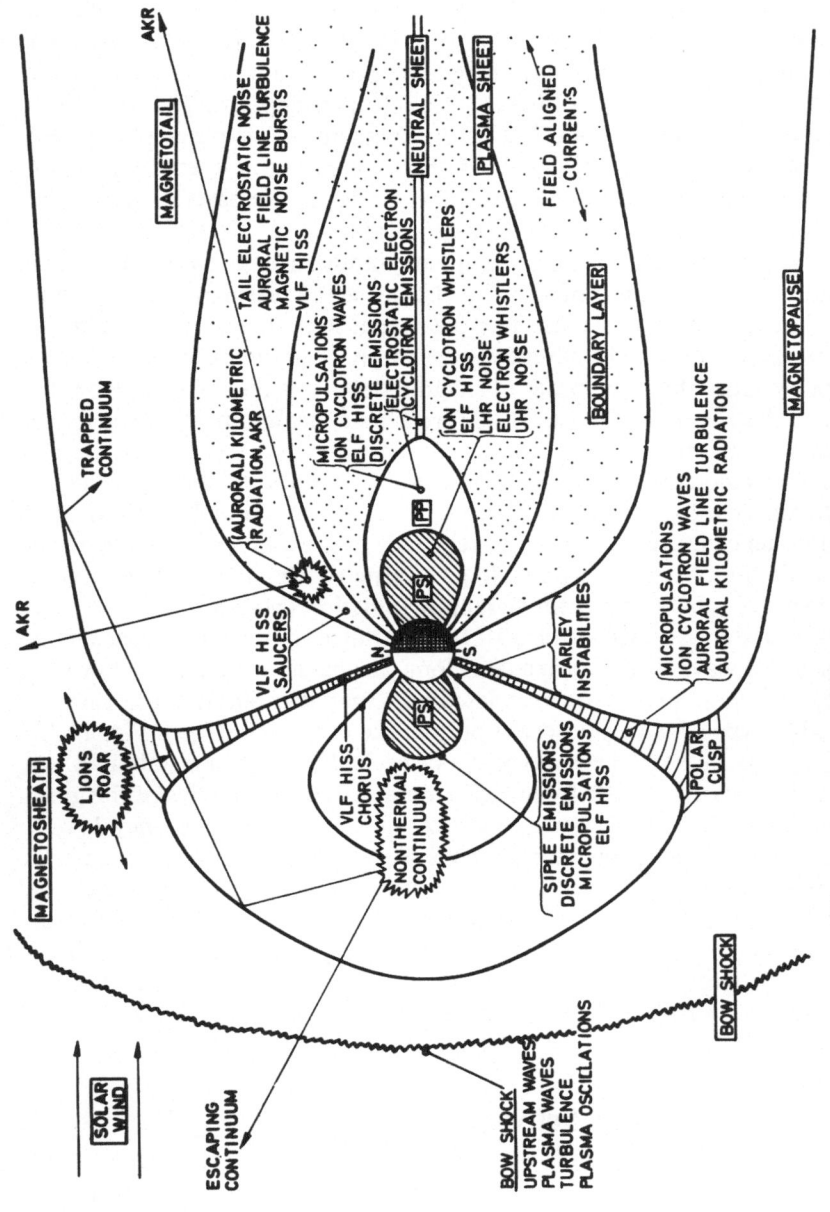

Fig. 13.11. Regions of plasma wave occurrence located in a noon—midnight meridional cross section of the magnetosphere. (Shawhan 1979)

The excitation mechanism of the discrete tones is very probably cyclotron resonance as discussed in Section 12.3 [Eq. (12.37)]. Rising and falling tones (Fig. 13.10) may be explained as arising from precipitating energetic electrons in a narrow energy range ($5-150$ keV) near the plasmapause. The location of excitation and the dispersion effect of the magnetospheric plasma are jointly responsible for the type of signal. If, for instance, the interaction region moves from the equator along a field line where f_{He} increases, a rising tone ($f \simeq f_{He}$) of the form of Fig. 13.10a is generated.

ELF hiss (<3000 Hz) with maximum amplitudes near a few hundred Hz is confined to a limited range of L shells just inside the plasmapause. It is most intense in the afternoon sector. The plasmapause is that surface where the value $B_0^2/(2\mu n_e)$ in Eq. (12.38) increases sharply because the electron density n_e drops by more than one order of magnitude as one moves away from the earth, and the energy $W_{\parallel}$ in Eq. (12.38) reaches a maximum at the equatorial plasmapause regions.

VLF hiss recorded on satellites at auroral field lines, mainly below about 2500 km altitude, has a sharp low frequency cutoff at about $5-10$ kHz. The cutoff frequency is the lower hybrid frequency of the ionosphere in the neighborhood of the observation point (e.g., Gurnett et al. 1983). Its excitation mechanism is probably coherent Cerenkov radiation (Maggs 1976).

Lion roars are signals in the $90-160$ Hz range, consisting of packets of whistler-mode waves which propagate along the magnetic field lines down the cusp region. These signals are associated with magnetospheric substorms.

Another type of wave, observed only in the cusp regions (see Fig. 13.11) are electrostatic waves. These waves can be excited by plasma instabilities in a hot turbulent plasma. Their frequencies are near half-harmonics of the gyro-frequency: $f \simeq (2\nu+1) f_{He}/2$ with $\nu = 1, 2, \ldots$. Waves of frequency $3f_{He}/2$ can become very strong between L shells 4 and 10. Their electric fields can reach 10 mV/m, which is well above the strength of the electric convection field in these regions [about $0.2-0.8$ mV/m at $L = 2-6$; (Maynard et al. 1983); see also Sect. 11.1]. Finally, proton heating in the cusp regions is associated with electrostatic ion-cyclotron waves at frequencies near the ion gyrofrequency. For more details, see the review by Shawhan (1979).

13.4 Man-Made Wave Activity

Electromagnetic signals transmitted from ground-based artificial radiators can propagate into the upper atmosphere and can interact with the ionospheric and magnetospheric plasma. Power lines, which radiate at 50 to 60 Hz and their higher harmonics, belong to the main sources of low frequency man-made noise. Most of this energy is trapped within the terrestrial wave guide between earth and ionosphere (see Sect. 7.3) and contributes to the ELF activity. The intensity of these ELF waves in industrial areas is much larger

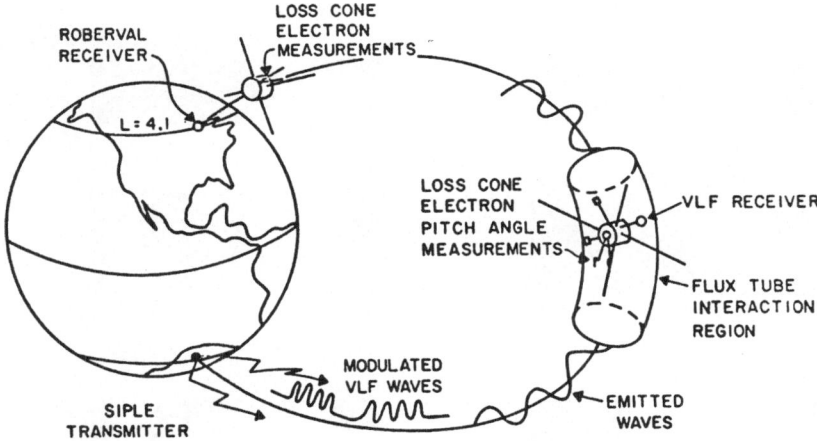

Fig. 13.12. Wave–particle interaction mechanism. Schematic illustration of wave injection from the Siple transmitter along the $L \simeq 4$ line toward the conjugate point at Roberval, Canada. These injected waves organize the energetic electrons to emit and amplify waves at adjacent but variable frequencies in the equatorial region. These large amplitude secondary waves cause electron precipitation. (Shawhan 1979)

then that of the lightning-induced Schumann resonances, discussed in Section 7.4.

Other artificial sources of ELF waves are, for instance, the SANGUINE ELF communication system, which worked in the frequency band between 40 and 80 Hz for a few years (e.g., Willim 1974), or *AC* currents at 16.67 Hz of the German railway system. A number of commercial VLF transmitters for communication purposes as well as navigation systems in the VLF range (OMEGA) and LF range (LORAN) are operating continuously (e.g., Swanson 1974). Furthermore, controlled ground-based VLF experiments are being conducted at Siple, Antarctica, to study the stimulation of whistler-mode waves and wave-particle interactions (e.g., Stiles 1974; Part et al. 1983).

The mechanism of wave–wave and wave–particle interaction is outlined in Fig. 13.12. Artificial VLF signals propagating in the whistler mode, but also naturally occurring whistlers, may amplify power line harmonic radiation (PLHR) in the VLF range by up to 30 dB and more during one passage through the interaction range, mainly in the equatorial region. This is a non-linear process resulting in frequency broadening. Without amplification, PLHR waves are usually below the threshold of detection. The injected waves may coherently drive energetic electrons to excite and amplify by cyclotron resonance free-running waves. The frequencies of these emissions are no longer controlled by the triggering waves. Figure 13.13 shows an example of rising tone emissions, starting at a high power line harmonic, which merge eventually to form broad noise bursts. PLHR waves also interact with other man-made noise or naturally occurring waves by entraining them or cutting

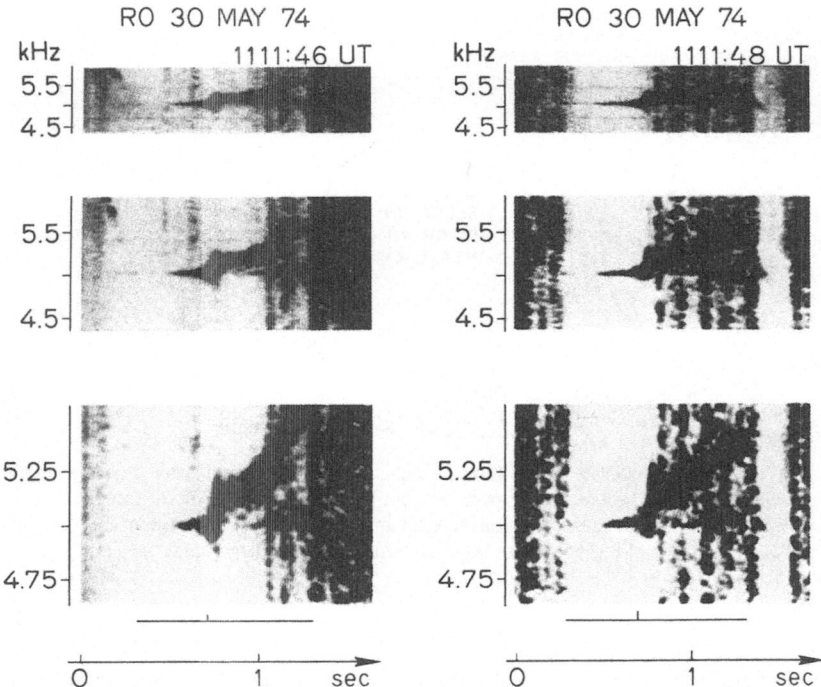

Fig. 13.13. Examples of typical gap-induced emission displayed at three different frequency scales (*upper, middle,* and *lower panel*). The triggering signal at 5 kHz has a pulse length of 1 s and is interrupted by a 10 ms gap after 400 ms (*horizontal line* at bottom with *small vertical bar*). A rising emission with $df/dt \simeq 8$ kHz/s is induced by each interruption. The emissions appear to feed energy to the 85-th and the 86-th power line harmonic of 60 Hz at 5.1 and 5.16 kHz, respectively, which in turn induce rising emissions with $df/dt \simeq 1$ kHz/s. (Chang and Helliwell 1979)

them off. Each time the PLHR waves pass through the interaction region, they may be further amplified or may trigger new emissions. Magnetospheric waves stimulated by PLHR waves therefore show a variety of spectral characteristics (Park et al. 1983). The PLHR effects measured at Siple, Antarctica, or at its conjugate point in Canada are observed between 1.5 and 8 kHz and are most intense near 3.5 kHz.

World maps of the occurrence of VLF emissions, obtained by satellite observations at midlatitudes, reveal maxima above industrial areas of high power consumption in North America, Europe, and Asia, and at their conjugate points in the southern hemisphere. This is indicative of the major role played by PLHR in stimulating these emissions. There appears to be also a correlation between the magnetospheric ring current (Sect. 11.6) and signal intensity of PLHR, suggesting the capability of turbulent energetic plasma in the equatorial region for wave amplification (Bullough 1983; Tatnall et al. 1983).

Finally, an experiment to locally heat the lower ionosphere with high-powered pulsed radio waves in the MHz range has been started recently at

Tromsø, Norway. Power modulation with periods in the $pc5$ range stimulates ionospheric AC currents which are superposed on the already existing DC currents at dynamo region heights (Stubbe and Kopka 1981). The magnetic effect of these stimulated AC currents has been observed on the ground as artificial magnetic pulsations with amplitudes up to a few nanotesla. Similarly, VLF waves have been stimulated by power modulation in that frequency range (Rietveld et al. 1983).

14 Appendix

14.1 Units and Coordinate Systems

Throughout this book, we use the International System of units (SI): meter kilogram, second, and ampere, or SI-derived units with prefixes ranging from pico (10^{-12}) to tera (10^{12}). Only occasionally are other units used, in particular within figures adopted from the literature. One main advantage of S: units is the equivalence between the units of mechanical and electric energy 1 joule = 1 watt $\times$ second.

Workers in the various scientific disciplines touched upon in this book unfortunately prefer different coordinate systems. The conventional coordinate system in meteorology is that of Fig. 14.1a with the x-axis directed to the east the y-axis directed to the north, and the z-axis directed upward. We usually apply this system. In the literature of atmospheric electricity, however, a coordinate system as in Fig. 14.1b is often adopted with the z-axis downward directed. A vertical electric field directed upward is defined as negative in this system. Several figures in Chapters 3 to 7, which are reproduced from the literature, use this coordinate system. In order to avoid confusion, the respective coordinate system applied is often specifically mentioned in the figure captions. Ionospheric physicists prefer the system of Fig. 14.1c with x directed to the south, y directed to the east, and z directed upward. This coordinate system fits naturally to the spherical coordinate system in Fig. 14.1d with radial distance r, colatitude θ (or latitude $\phi = \pi/2 - \theta$), and longitude λ.

A geomagnetic spherical coordinate system with the dipole axis as polar axis is often used in geomagnetic studies (Fig. 14.1e) with (θ_p, λ_p) the geographic coordinates of the dipole axis in the northern hemisphere [see Eq (14.5)], (θ, λ) the geographic coordinates of a location P on the sphere, and (θ_m, λ_m) its geomagnetic coordinates. Transformation from geographic to geomagnetic coordinates is a simple exercise in spherical trigonometry. A somewhat more complicated coordinate system uses invariant latitude and longitude (or magnetic local time) based on the realistic geomagnetic field configuration on the ground, instead of the geomagnetic coordinates (e.g.; Akasofu and Chapman 1972). For our purpose, the difference between invariant and geomagnetic coordinates is irrelevant.

Magnetospheric physicists, finally, often apply the solar-magnetospheric coordinate system of Fig. 14.1f with the x-axis pointed to the sun, the z-axis upward directed orthogonal to the x-axis chosen such that the $x - z$ plane in-

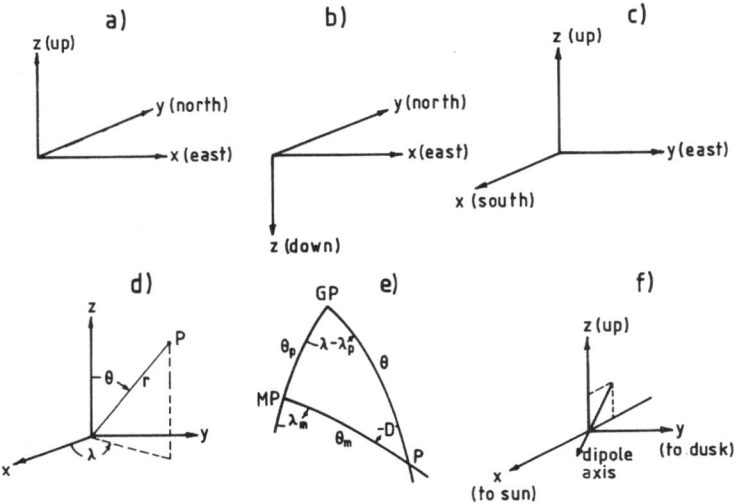

Fig. 14.1a–f. Various coordinate systems used throughout this book. **a** Meteorological system (this is the basic coordinate system applied in the book). **b** System preferred by workers in the field of atmospheric electricity. **c** System preferred by ionospheric physicists. **d** Spherical coordinate system. **e** Geomagnetic coordinates (*GP*: geographic pole; *MP*: geomagnetic pole). **f** Solar-magnetospheric coordinate system

cludes the geomagnetic dipole axis, and the *y*-axis in the equatorial plane pointed to the dusk side so that the coordinate system is right handed (Akasofu and Chapman 1972). For our purpose, it is often sufficient to neglect the wobble of the geomagnetic dipole axis around the geographic axis and consider the dipole axis parallel to the *z*-axis (co-axial dipole).

14.2 Geomagnetic Dipole

The main geomagnetic field is of internal origin and is generated within the outer core of the earth. Its manifestation on the ground, to a first approximation, is the field of an inclined magnetic dipole located at the earth's center with its axis directed to the south. Only about 10% of the observed geomagnetic field on the ground is a nondipole field (e.g., Merrill and McElhinny 1983). Secular variations are slow changes of the internal field with time scales ranging from years to millions of years. For the time scales considered in this book (less than 22 years), these secular variations are of minor importance, and the internal field is treated as a time invariant field. Furthermore, we ignore the nondipole component.

The magnetic field components of the geomagnetic dipole with its north pole in the southern hemisphere can be derived from

$$B = -\nabla \Phi, \tag{14.1}$$

where the scalar potential (in the geomagnetic coordinate system of Fig. 14.1e) is given by

$$\Phi = -\mu_0 M \cos \theta_m / (4\pi r^2), \tag{14.2}$$

with $\mu_0 = 4\pi \times 10^{-7}$ H/m the permeability of free space, and $M = 7.91 \times 10^{22}$ Am^2 the magnetic moment of the geomagnetic dipole. The components of the magnetic field are

$$B_r = -2B_{00} \cos \theta_m (a/r)^3$$
$$B_\theta = -B_{00} \sin \theta_m (a/r)^3, \tag{14.3}$$

with $a = 6371$ km the earth's radius, and

$$B_{00} = \mu_0 M / (4\pi a^3) = 3.06 \times 10^{-5}\, \text{T} \tag{14.4}$$

the dipole field strength at the magnetic equator on the ground.

Presently (epoch of 1980), the south pole of the geomagnetic dipole is located at

$$\theta_p = 11.20°; \quad \lambda_p = 70.75° \, \text{W}. \tag{14.5}$$

The horizontal component, directed to magnetic north, ($H = -B_\theta$; see Fig. 14.2), is called the horizontal intensity. The vertical component, directed downward ($Z = -B_r$), is called the vertical intensity, and the angle between geographic and geomagnetic north (measured in clockwise direction; see Fig. 14.1e), is called the declination D. The angle between H and Z is the dip angle I, and the total intensity is

$$F = (H^2 + Z^2)^{1/2} = B_{00}(1 + 3\cos^2 \theta_m)^{1/2}(a/r)^3. \tag{14.6}$$

The dip angle is related to the geomagnetic latitude by

$$\tan I = Z/H = 2\tan \phi_m. \tag{14.7}$$

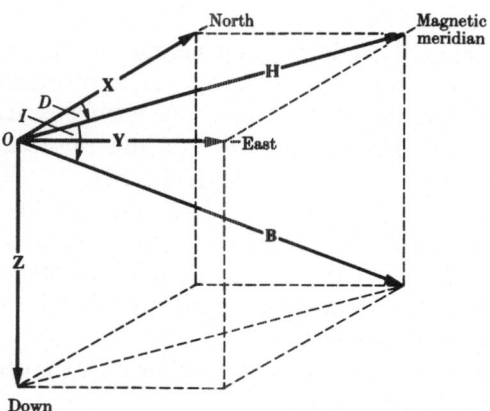

Fig. 14.2. Definition of geomagnetic field components

The (geographic) north component X and the (geographic) east component Y are related to H and D by

$$X = H \cos D; \quad Y = H \sin D. \tag{14.8}$$

The variations of the geomagnetic field due to external sources are decomposed into $(\Delta H, \Delta D, \Delta Z)$ or $(\Delta X, \Delta Y, \Delta Z)$, respectively. These are small deviations from the main field on the earth's surface ($<1\%$), but may be of the same order of magnitude or larger at distances beyond a few earth radii.

A useful parameter is the L shell parameter which describes the course of a single geomagnetic field line. In the case of a dipole, L is defined by

$$L = r/(a \sin^2 \theta_m) \tag{14.9}$$

with $\theta_0 = \arcsin(1/\sqrt{L})$ the colatitude of the foot print of the line on the ground ($r = a$), and La the radial distance of the line at the geomagnetic equator ($\theta_m = \pi/2$). The length of a dipole line outside the earth's surface is given by

$$S = 2La \int_{\theta_0}^{\pi/2} (1 + 3 \cos^2 \theta)^{1/2} \sin \theta \, d\theta$$
$$= La\{x(1+3x^2)^{1/2} + \ln[\sqrt{3}x + (1+3x^2)^{1/2}]/\sqrt{3}\} \approx a(2.76L - 2), \tag{14.10}$$

with $x = \cos \theta_0 = (1 - 1/L)^{1/2}$, and the last approximation valid for $L > 2$.

In this book, we symbolize the dipole field by B_0, as compared with a field of external origin, which is designed by B.

14.3 Basic Formulas of Electromagnetism and Hydromagnetics

Maxwell's equations are (e.g., Menzel 1960)

$$\nabla \times (B/\mu) = \varepsilon \partial E/\partial t + j \quad \text{(Ampere's law)} \tag{14.11}$$

$$\nabla \times E = -\partial B/\partial t \quad \text{(Faraday's law)} \tag{14.12}$$

$$\nabla \cdot (\varepsilon E) = q \quad \text{(Coulomb's law)} \tag{14.13}$$

$$\nabla \cdot B = 0 \quad \text{(absence of magnetic poles)}, \tag{14.14}$$

with

B magnetic induction (in T) (in this book, referred to as magnetic field)
E electric field (in V/m)
q electric charge density (in C/m^3)
j electric current density (in A/m^2)
ε electric permittivity (in F/m)
μ permeability (in H/m).

Throughout this book, we assume

$$\mu \simeq \mu_0 = 4\pi \times 10^{-7}\,\text{H/m}; \quad \varepsilon \simeq \varepsilon_0 = 8.854 \times 10^{-12}\,\text{F/m}, \tag{14.15}$$

where μ_0 and ε_0 are the respective values in vacuo. They are related to the speed of light c by

$$c \simeq 1/(\varepsilon_0\mu_0)^{1/2}\,\text{m/s}. \tag{14.16}$$

Ohm's law in general form yields

$$j = \sigma \cdot (E + v \times B), \tag{14.17}$$

with σ an electric conductivity tensor [in S/m; note that siemens (S) = mhos $(1/\Omega)$], and v a (nonrelativistic) velocity of the plasma. Combining Eqs. (14.11) and (14.13) yields a continuity equation for the current:

$$\partial q/\partial t + \nabla \cdot j = 0. \tag{14.18}$$

In many applications of atmospheric electrodynamics, the displacement current $\varepsilon\partial E/\partial t$ in Eq. (14.11) can be neglected. This is allowed if the characteristic time τ of a disturbance is much larger than $2\pi\varepsilon/\sigma$. Within the magnetosphere, the characteristic time where displacement currents can be neglected is larger than a fraction of a second. Within the lower atmosphere, τ must be larger than several minutes.

If the displacement current is neglected and if the conductivity σ is a constant scalar, a hydromagnetic equation can be derived from Eqs. (14.11), (14.12) and (14.17):

$$\partial B/\partial t = \nabla \times (v \times B) - 1/(\mu\sigma)\nabla \times (\nabla \times B). \tag{14.19}$$

For large values of σ, the second term on the right hand side in Eq. (14.19) is small, and the magnetic lines of force are considered as frozen into the plasma (Alfven and Fälthammar 1963). This corresponds to zero current density in Eq. (14.17), and thus to the condition

$$E = -v \times B. \tag{14.20}$$

If the displacement current is neglected, the electric current density in Eq. (14.11) is directly related to the magnetic field configuration via the Biot-Savart law:

$$B(r) = (\mu/4\pi)\int_V j(r') \times R\,dV/R^3, \tag{14.21}$$

where $R = r - r'$ is the distance between the integration point at r' within the volume V where the current flows, and the location r of the observer.

Some simple current configurations are shown in Fig. 14.3. The magnetic field along the central axis of the circular current loop of radius R and current strength J_λ in Fig. 14.3a is

$$B_z = \mu J_\lambda R^2/[2(R^2 + z^2)^{3/2}]. \tag{14.22}$$

At great distances, the magnetic field of the loop current equals that of a dipole with magnetic moment

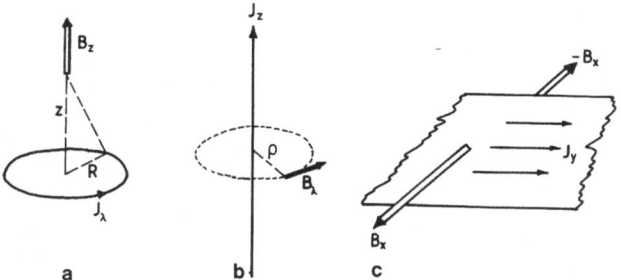

Fig. 14.3a–c. Electric currents and their magnetic field configurations. **a** Current loop. **b** Line current. **c** Sheet current

$$M = J_\lambda \pi R^2 . \tag{14.23}$$

The azimuthal magnetic field of a current within an infinitely extended straight wire in Fig. 14.3b is

$$B_\lambda = \mu J_z / (2\pi \varrho) . \tag{14.24}$$

The magnetic field above and below an infinitely extended current sheet of thickness Δh and current strength per unit length $J_y = j_y \Delta h$, flowing in the y-direction (Fig. 14.3c), is

$$B_x = \pm \mu J_y / 2 . \tag{14.25}$$

The magnetic field reverses its direction between top and bottom of the sheet current.

We can determine the direction of the magnetic field with respect to the current direction from the simple "rule of thumb": if the thumb of the right hand points in the direction of the current, the curved fingers give the direction of the associated magnetic field.

We often use in this book electric fields which are derived from an electrostatic potential. This is possible if the left hand side in Eq. (14.19) can be neglected, or if the characteristic time of a disturbance is much larger than $\sigma \mu h^2$ where h is a characteristic length. For global scale atmospheric dynamics ($h \simeq a$), this characteristic time is larger than about 1 h.

14.4 List of Frequently Used Symbols

$a = 6371.2$ km   Mean earth's radius
$B = (B_x, B_y, B_z)$ Magnetic induction (referred to as magnetic field in this book)
B_0 Geomagnetic (dipole) field
$B_{00} = 30572$ nT   Geomagnetic dipole field strength at the equator on the ground (epoch 1980)

C	Capacitance
$c = 2.998 \times 10^8 \, \text{m/s}$	Speed of light
$c_p, \, c_v$	Specific heats at constant pressure and volume, respectively
$\boldsymbol{E} = (E_x, \, E_y, \, E_z)$	Electric field
$e = 1.602 \times 10^{-19} \, \text{C}$	Elementary electric charge
ε	Permittivity
$\varepsilon_0 = 8.854 \times 10^{-12} \, \text{F/m}$	Permittivity of free space
F	Area
$F_E = 5.101 \times 10^{14} \, \text{m}^2$	Surface area of earth
f	Frequency
Φ	Electric potential
ϕ	Latitude
ϕ_m	Geomagnetic latitude
G	Conductance
$g = 9.807 \, \text{m/s}^2$	Mean gravitational acceleration on the ground
Γ	Columnar resistance (also: lapse rate)
H	Scale height
h	Vertical distance (also: equivalent depth)
I	Dip angle of geomagnetic field
$i = (-1)^{1/2}$	Imaginary unit
$\boldsymbol{J} = (J_x, \, J_y, \, J_z)$	Electric current
$\boldsymbol{j} = (j_x, \, j_y, \, j_z)$	Electric current density
K	Normalized wave number
$k = \omega/c$	Wave number in vacuo (also: mobility factor)
L	Inductance
λ	Longitude (also: wave length)
M	Electric dipole moment (also: magnetic dipole moment)
m	Zonal wave number
$m_i, \, m_e$	Mass of ions and electrons, respectively
μ	Permeability
$\mu_0 = 4\pi \times 10^{-7} \, \text{H/m}$	Permeability of free space (throughout this book, $\mu \simeq \mu_0$ is assumed)
N	Total number of events (particles, thunderstorms, etc.)
n	Refractive index (also: meridional wave number)
$n_i, \, n_e$	Number density of ions and electrons, respectively
ν	Collision frequency (also: wave mode number)
$\Omega = 7.292 \times 10^{-5} \, \text{s}^{-1}$	Angular frequency of one siderial day
$\omega = 2\pi f$	Angular frequency
$\omega_i, \, \omega_e$	Plasma frequency of ions and electrons, respectively
$\omega_{Hi}, \, \omega_{He}$	Gyrofrequency of ions and electrons, respectively
Ψ	Stream function

P	Power
p	Pressure
Π	Ion production rate
Q	Electric charge
q	Electric charge density
R	Resistance (also: radius of sphere; also: reflection factor)
r	Radial distance
$\hat{r}$	Unit vector in radial direction
ϱ	Specific resistance (also: horizontal distance; also: density)
Σ	Height-integrated electric conductivity
σ	Electric conductivity
$\boldsymbol{\sigma}$	Conductivity tensor
σ_p, σ_h, σ_c, σ_F	Pedersen-, Hall-, Cowling-, Parallel-conductivity
T	Temperature (also: transmission function)
t	Time
τ	Local time (also: characteristic time)
θ	Colatitude
U	Voltage
$U = (u, v)$	Horizontal wind
V	Volume
$v = (u, v, w)$	Wind vector
v_D	Drift velocity
v_g	Group velocity
v_p	Phase velocity
W	Energy
w	Vertical wind (also: energy density)
(x, y)	Horizontal coordinates (see Sect. 14.1)
z	Vertical coordinate

References

Akasofu SI (1977) Physics of Magnetospheric Substorms. Reidel, Dordrecht, Holland
Akasofu SI (1980) In Akasofu SI (ed) Dynamics of the Magnetosphere. Reidel, Dordrecht, Holland, p 447
Akasofu SI (1981) Space Sci Rev 28:121
Akasofu SI, Chapman S (1972) Solar-Terrestrial Physics. Clarendon Press, Oxford
Alfven H, Fälthammar CG (1963) Cosmical Electrodynamics. Clarendon Press, Oxford
Anderson RV (1977) In Dolezalek H, Reiter R (eds) Electrical Processes in Atmospheres. Steinkopff, Darmstadt, p 87
Andrews MK, et al (1981) In Southwood DJ (ed) ULF Pulsations in the Magnetosphere. Reidel, Dordrecht, Holland, p 141
Arnold F (1980) In ESA Symposium on Rockets and Balloon Programs. Bournemouth, ESA, Paris, p 479
Aufdermauer AN, Johnson DA (1972) QJR Meteorol Soc 98:369
Axford WI (1969) Rev Geophys 7:421
Banks PM (1979) In Kennel CF, et al (eds) Solar System Plasma Physics, Vol II North Holland, Amsterdam, p 57
Barbosa DD (1982) Rev Geophys Space Phys 20:316
Barry JD (1980) Ball Lightning and Bead Lightning. Plenum Press, New York
Baum CE, et al (1978) IEEE Transact Antennas Progag AP-26:22
Baumjohann W (1983) Adv Space Res II 10:55
Baumjohann W, Glassmeier K (1984) Planet Space Sci 32 (in press)
Baumjohann W, et al (1980) J Geophys Res 85:1963
Berger K (1977) In Golde RH (ed) Lightning, Vol I. Academic Press, London, New York, p 119
Blanc M, Richmond AD (1980) J Geophys Res 85:1669
Boller BR, Stolov HL (1970) J Geophys Res 75:6073
Boström R (1964) J Geophys Res 69:4883
Boström R, Fahleson U (1977) In Dolezalek H, Reiter R (eds) Electrical Processes in Atmospheres. Steinkopff, Darmstadt, p 529
Brook M, Ogawa T (1977) In Golde RH (ed) Lightning, Vol I. Academic Press, London, New York, p 191
Brown GM (1975) J Atmos Terr Phys 37:107
Bullough K (1983) Space Sci Rev 35:175
Burke WJ (1982) Rev Geophys Space Phys 20:685
Burrows K, Hall SH (1965) J Geophys Res 70:2149
Campbell WH (1967) In Matsushita S,Campbell WH (eds) Physics of Geomagnetic Phenomena, Vol II. Academic Press, London, New York, p 821
Campbell WH, Matsushita S (1982) J Geophys Res 87:5305
Cap FF (1976) Handbook of Plasma Instabilities, Vol I. Academic Press, London, New York
Carpenter DL, Park CG (1973) Rev Geophys Space Phys 11:133
Chalmers JA (1967) Atmospheric Electricity. Pergamon Press, Oxford
Chang DCD, Helliwell RA (1979) J Geophys Res 84:7170
Chapman S, Bartels J (1951) Geomagnetism. Clarendon Press, Oxford
Chapman S, Lindzen RS (1970) Atmospheric Tides. Reidel, Dordrecht, Holland
Chappel CR (1972) Rev Geophys Space Phys 10:951

Chernosky EJ (1966) J Geophys Res 71:965
Clauer CR, et al (1983) J Geophys Res 88:2123
Cole KD (1974) Planet Space Sci 22:1075
Crochet M, et al (1979) J Geophys Res 84:5223
Davis TN, et al (1965) J Geophys Res 70:5883
Davis TN, et al (1967) J Geophys Res 72:1845
Denisse JF, Delcroix JL (1963) Plasma Waves. Wiley Interscience, New York
Dennis AS, Pierce ET (1964) Radio Sci 68D:777
Dibner B (1977) In Golde RH (ed) Lightning, Vol I. Academic Press, London, New York, p 23
Dolezalek H (1972) Pure Appl Geophys 100:8
Dolezalek H, Reiter R (eds) (1977) Electrical Processes in Atmospheres. Steinkopff, Darmstadt
Dungey JW (1967) In Matsushita S, Campbell WH (eds) Physics of Geomagnetic Phenomena, Vol I. Academic Press, London, New York, p 913
Eastman TE, et al (1976) Geophys Res Lett 3:685
Edgar RC (1976) J Geophys Res 81:205
Evans JV (1978) Rev Geophys Space Phys 16:195
Fejer JA (1964) Rev Geophys 2:275
Fejer BG, Kelly MC (1980) Rev Geophys Space Phys 18:401
Few AA (1982) In Volland H (ed) Handbook of Atmospherics, Vol II. CRC Press, Boca Raton, Fla, p 57
Feynman J (1983) Rev Geophys Space Phys 21:338
Fischer HJ (1977) Prometheus 7:4
Forbes JM (1981) Rev Geophys Space Phys 19:469
Forbes JM (1982a) J Geophys Res 87:5222
Forbes JM (1982b) J Geophys Res 87:5241
Forbes JM, Lindzen RS (1976a) J Atmos Terr Phys 38:897
Forbes JM, Lindzen RS (1976b) J Atmos Terr Phys 38:911
Fukushima N (1971) Radio Sci 6:269
Gendrin R (1981) Rev Geophys Space Phys 19:171
Gendrin R (1983) Space Sci Rev 34:271
Gibson EG (1973) The Quiet Sun. NASA SP-303. Scientific and Technical Information Office, Washington, DC
Golde RH (1977) In Golde RH (ed) Lightning, Vol I. Academic Press, London, New York, p 309
Greifinger C, Greifinger PS (1965) J Geophys Res 70:2217
Grenet G (1947) Extr An Geophys 3:306
Gringel W (1977) Prometheus 2:13
Gringel W, et al (1978) Meteor Forsch Ergebn. B13:41
Gurevich AV, et al (1976) Space Sci Rev 19:59
Gurnett DA (1965) In McCormac BM (ed) Magnetospheric Particles and Fields. Reidel, Dordrecht, Holland, p 197
Gurnett DA, et al (1976) J Geophys Res 70:1665
Gurnett DA, et al (1983) J Geophys Res 88:329
Haerendel G, Paschmann G (1975) In Hultquist B, Stenflo L (eds) Physics of the Hot Plasma in the Magnetosphere. Plenum Press, New York, p 23
Hanuise C, et al (1983) J Geophys Res 88:253
Harel M, et al (1981) J Geophys Res 86:2217
Harth W (1982) In Volland H (ed) Handbook of Atmospherics, Vol II. CRC Press, Boca Ratin, Fla, p 133
Hays PB, Roble RG (1979) J Geophys Res 84:3291
Heacock HH, Hunsucker RD (1981) Space Sci Rev 28:191
Heelis RA, et al (1982) J Geophys Res 87:6339
Helliwell RA (1965) Whistlers and Related Ionospheric Phenomena. Univ Press, Stanford
Heppner JP (1972) In Dyer ER (ed) Critical Problems of Magnetospheric Physics. Natl Acad Sci, Washington DC, p 107

Herman JR, Goldberg RA (1978) Sun, Weather, and Climate. NASA SP-426, Scientific and
    Technical Information Branch, Washington, DC
Hill RD (1963) J Geophys Res 71:1966
Hill RD (1977) In Golde RD (ed) Lightning, Vol I. Academic Press, London, New York, p 385
Hill RD (1979) Rev Geophys Space Phys 17:155
Hill TW, Rassbach ME (1975) J Geophys Res 80:1
Hoeksema JT, et al (1983) J Geophys Res 88:9910
Hofmann DJ, Rosen JM (1977) J Geophys Res 82:1435
Hones EW (1979a) In McCormack BM, Seliga TA (eds) Solar Terrestrial Influences on Weather
    and Climate. Reidel, Dordrecht, Holland, p 83
Hones EW (1979b) Space Sci Rev 23:393
Hughes WJ, Southwood DJ (1976) J Geophys Res 81:3234
Hundhausen AJ (1979) Rev Geophys Space Phys 17:2034
Iijima T, Potemra TA (1978) J Geophys Res 83:599
Iribarne JV, Cho HR (1980) Atmospheric Physics. Reidel, Dordrecht, Holland
Israel H (1970) Atmospheric Electricity, Vol I. Natl Sci Found, Washington, DC
Israel H (1973) Atmospheric Electricity, Vol II. Natl Sci Found, Washington, DC
Kamide Y, Baumjohann W (1984) J Geophys Res 89 (in press)
Kamide Y, Matsushita S (1979) J Geophys Res 84:4083
Kamide Y, et al (1981) J Geophys Res 86:801
Kamide Y, et al (1982) J Geophys Res 87:8228
Kasemir H (1959) Z Geophys 25:33
Kato S (1956) Geomagn Geoelectr 8:24
Kato S (1980) Dynamics of the Upper Atmosphere. Reidel, Dordrecht, Holland
Kelley MC, et al (1983) Geophys Res Lett 10:733
Kennel CF, et al (eds) (1979) Solar System Plasma Physics, Vols II, III. North-Holland,
    Amsterdam
Kertz W (1969) Einführung in die Geophysik, Vol II. BI Hochschultaschenbücher, Mannheim
Kivelson MG (1976) Rev Geophys Space Phys 14:189
Kivelson MG, et al (1980) In Akasofu IS (ed) Dynamics of the Magnetosphere. Reidel, Dordrecht,
    Holland, p 385
Krider EP, Musser JA (1982) J Geophys Res 87:11171
Krider EP, et al (1977) J Geophys Res 82:951
Kuettner JP, et al (1981) J Atmos Sci 38:2470
Lanzerotti LJ, Southwood DJ (1979) In Lanzerotti LJ, et al (eds) Solar System Plasma Physics,
    Vol III. North-Holland, Amsterdam, p 109
Latham J, Stromberg IM (1977) In Golde RH (ed) Lightning, Vol I. Academic Press, London,
    New York, p 54
Leise JA, Taylor WL (1977) J Geophys Res 82:391
Lewis EA (1982) In Volland H (ed) Handbook of Atmospherics, Vol I. CRC Press, Boca Raton,
    Fla, p 253
Lilly DK (1979) Annu Rev Earth Planet Sci 7:117
Lin YT, et al (1979) J Geophys Res 84:6307
Lin YT, et al (1980) J Geophys Res 85:1571
Longmire CL (1978) IEEE Trans Antennas and Propagation AP-26:3
MacGorman DR, et al (1981) J Geophys Res 86:9900
Maeda H (1974) J Atmos Terr Phys 36:1395
Maeda H, Kamei T (1975) Report Ionos Space Res Jpn 29:177
Maeda H, et al (1982) Geophys Res Lett 9:337
Maezawa K (1974) Planet Space Sci 22:1443
Maggs JE (1976) J Geophys Res 81:1707
Magid LM (1972) Electromagnetic Fields, Energy, and Waves. Wiley & Sons, New York
Magono C (1980) Thunderstorms. Elsevier, Amsterdam
Malin SRC (1973) Philos Trans R Soc London Ser A 274:551
Mansurov SM (1969) Geomagn Aeron 9:622

Marshall JS, Palmer WM (1948) J Meteorol 5:165
Markson R (1981) Nature (London) 291:304
Mason BJ (1971) The Physics of Clouds. Clarendon Press, Oxford
Mathal KC, et al (1980) Rev Geophys Space Phys 18:361
Matsushita S (1967) In Matsushita S, Campbell WH (eds) Physics of Geomagnetic Phenomena, Vol I. Academic Press, London, New York, p 301
Matsushita S, Campbell WH (eds) (1967) Physics of Geomagnetic Phenomena, Vols I, II. Academic Press, London, New York
Mayaud PN (1980) Derivation, Meaning, and Use of Geomagnetic Indices, Geophys Monogr 22, AGA, Washington DC
Maynard NC (1974) J Geophys Res 79:4620
Maynard NC, et al (1981) Geophys Res Lett 8:923
Maynard NC, et al (1983) J Geophys Res 88:3991
McDonald TB, et al (1979) J Geophys Res 84:1727
Mendillo M, Papagiannis MD (1971) J Geophys Res 76:6939
Menzel DH (ed) (1960) Fundamental Formulas of Physics, Vol I. Dover, New York
Merrill RT, McElhinny MW (1983) The Earth's Magnetic Field. Academic Press, London, New York
Meyerott RE, et al (1983) In McCormack BM (ed) Weather and Climatic Responses to Solar Variation. Colorado Assoc Univ Press, Boulder
Möhlmann D (1974) Gerland's Beitr Geophys 83:16
Möhlmann D (1977) J Atmos Terr Phys 39:1325
Mohnen VA (1977) In Dolezalek H, Reiter R (eds) Electrical Processes in Atmospheres. Steinkopff, Darmstadt, p 1
Moore CB, Vonnegut B (1977) In Golde RH (ed) Lightning, Vol I. Academic Press, London, New York, p 51
Mozer FS (1970) Planet Space Sci 18:259
Mozer FS (1981) In Akasofu JI, Kan SR (eds) Physics of Auroral Arc Formation. Geophys Monogr 25, AGU, Washington, DC, p 139
Mühleisen R (1971) Z Geophys 37:759
Murata H (1974) Space Sci Rev 16:461
Nagano H, et al (1981) Planet Space Sci 29:529
Nisbet JS, et al (1978) J Geophys Res 83:2647
Nishida A (1978) Geomagnetic Diagnosis of the Magnetosphere. Springer, Berlin, Heidelberg, New York
Nishida A, Kokubun S (1971) Rev Geophys Space Phys 9:417
Oetzel GN (1968) J Geophys Res 73:1889
Ogawa T (1982) In Volland H (ed) Handbook of Atmospherics, Vol I. CRC Press, Boca Raton, Fla, p 24
Ogawa T, et al (1969) J Geomagn Geoelectr 21:447
Onwumechilli A (1967) In Matsushita S, Campbell WH (eds) Physics of Geomagnetic Phenomena, Vol I. Academic Press, London, New York, p 426
Orville RE (1977) In Golde RH (ed) Lightning, Vol I. Academic Press, London, New York, p 281
Orville RE (1982) In Volland H (ed) Handbook of Atmospherics, Vol II. CRC Press, Boca Raton, Fla, p 79
Park CG (1976) J Geophys Res 81:168
Park CG (1982) In Volland H (ed) Handbook of Atmospherics, Vol II. CRC Press, Boca Raton, Fla, p 21
Park CG, Dejnakarintra M (1973) J Geophys Res 78:6623
Park CG, et al (1983) Space Sci Rev 35:131
Pathak PP, et al (1980) Ann Geophys 36:613
Pathak PP, et al (1982) Geophys J R Astron Soc 69:197
Pierce ET (1977) In Golde RG (ed) Lightning, Vol I. Academic Press, London, New York, p 351
Polk C (1982) In Volland H (ed) Handbook of Atmospherics, Vol I. CRC Press, Boca Raton, Fla, p 112

Prentice SA (1977) In Golde RH (ed) Lightning, Vol I. Academic Press, London, New York, p 465

Price AT (1967) In Matsushita S, Campbell WH (eds) Physics of Geomagnetic Phenomena, Vol I. Academic Press, London, New York, p 235

Price GH (1974) Rev Geophys Space Phys 12:389

Price GH, Pierce ET (1977) Radio Sci 12:381

Pruppacher MR, Klett JD (1980) Microphysics of Clouds and Precipitation. Reidel, Dordrecht, Holland

Pudovkin MI (1974) Space Sci Rev 16:727

Ratcliffe JR (1972) An Introduction to the Ionosphere and Magnetosphere. Univ Press, Cambridge

Reagan JB, et al (1983) J Geophys Res 88:3869

Rees MH, Roble RG (1975) Rev Geophys Space Phys 13:201

Reiff PH, et al (1981) J Geophys Res 86:7639

Reiter R (1980) Bauphysik 2:88

Richmond AD (1973a) J Atmos Terr Phys 35:1083

Richmond AD (1973b) J Atmos Terr Phys 35:1105

Richmond AD (1976) J Geophys Res 81:1447

Richmond AD, Venkateswaran SV (1971) Radio Sci 6:139

Richmond AD, et al (1976) J Geophys Res 81:547

Rietveld MT, et al (1983) J Geophys Res 88:2140

Roble RG, Hays PB (1979) J Geophys Res 84:7247

Roederer JG (1970) Dynamics of Geomagnetically Trapped Radiation. Springer, Berlin, Heidelberg, New York

Rosenbauer H, et al (1975) J Geophys Res 80:2723

Rosenberg TJ, et al (1981) Phys Rev Lett 47:1343

Rostoker G (1972) Rev Geophys Space Phys 10:935

Rostoker G (1979) Fund Cosm Phys 4:211

Rostoker G (1983) In Hultquist B, Hagfors T (eds) High-Latitude Space Plasma Physics. Plenum Press, New York, London, p 189

Rostoker G, Boström R (1976) J Geophys Res 81:235

Russell CT (1974) Geophys Res Lett 1:11

Rycroft MJ (1974) In Holtet JA (ed) ELF-VLF Radio Wave Propagation. Reidel, Dordrecht, Holland, p 317

Rycroft MJ, Thomas JO (1970) Planet Space Sci 18:65

Saflekos NA, et al (1982) Rev Geophys Space Phys 20:709

Saito T (1978) Space Sci Rev 21:427

Salah JE, Evans JV (1977) J Geophys Res 82:2413

Salanave LE (1980) Lightning and its Spectrum. Univ Press, Tuscon, Ariz

Sao K, Jindoh H (1974) J Atmos Terr Phys 36:261

Scarf FL (1975) In Hultquist B, Stenflo L (eds) Physics of the Hot Plasma in the Magnetosphere. Plenum Press, New York, London, p 371

Schindler K (1980) In Akasofu SI (ed) Dynamics of the Magnetosphere. Reidel, Dordrecht, Holland, p 311

Schulz M (1976) In Williams DJ (ed) Physics of Solar Planetary Environment, Vol I. AGU, Washington, DC, p 491

Schulz M, Lanzerotti LJ (1974) Particle Diffusion in the Radiation Belt. Springer, Berlin, Heidelberg, New York

Sckopke N (1972) Cosm Electrodyn 3:330

Shawhan SD (1979) In Lanzerotti LJ, et al (eds) Solar System Plasma Physics. Vol III. North-Holland, Amsterdam, p 211

Siebert M (1971) In Flügge S (ed) Handbuch der Physik, 49(3). Springer, Berlin, Heidelberg, New York, p 206

Siebert M (1977) Kleinheubacher Ber 21:313

Siscoe GI (1979) Planet Space Sci 27:287

Siscoe GI (1982) J Geophys Res 87:5124
Smith EJ (1979) Rev Geophys Space Phys 17:610
Sommerfeld A (1952) Lectures in Theoretical Physics, Vol III. Academic Press, London, New York
Sonnerup BU (1979) In Lanzerotti LJ, et al (eds) Solar System Plasma Physics, Vol III. North-Holland, Amsterdam, p 45
Southwood DJ (1981) In Southwood DJ (ed) ULF Pulsations in the Magnetosphere. Reidel, Dordrecht, Holland, p 75
Southwood DJ, Hughes WJ (1983) Space Sci Rev 35:301
Spaulding AD (1982) In Volland H (ed) Handbook of Atmospherics, Vol I. CRC Press, Boca Raton, Fla, p 289
Spiro RW, et al (1981) J Geophys Res 86:2261
Stening RJ (1969) Planet Space Sci 17:889
Stening RJ (1973) Planet Space Sci 21:1897
Stern DP (1973) J Geophys Res 78:7292
Stern DP (1983) Rev Geophys Space Phys 21:125
Stiles GS (1974) In Holtet JA (ed) ELF-VLF Radio Wave Propagation. Reidel, Dordrecht, Holland, p 335
Stubbe P, Kopka HJ (1981) J Geophys Res 86:1606
Sturrock PA (ed) (1980) Solar Flares. Col Assoc Univ Press, Boulder, Col
Sugiura M, Chapman S (1960) Abh Akad Wiss Göttingen, Math-Phys Kl, Spec Iss 4:53
Svalgaard L (1969) Geophys Pap R-6, Dan Meteor Inst, Copenhagen
Svalgaard L (1973) J Geophys Res 78:2064
Svalgaard L (1975) SUIPR Report No 646, Inst Plasma Res, Univ Press, Stanford, Cal
Svalgaard L (1977) In Zirker JB (ed) Coronal Holes and High Speed Wind Streams. Col Assoc Press, Boulder, Col, p 371
Swanson ER (1974) In Holtet JA (ed) ELF-VLF Radio Wave Propagation. Reidel, Dordrecht, Holland, p 371
Takeda M, Maeda H (1980) J Geophys Res 85:6895
Takeda M, Maeda H (1981) J Geophys Res 86:5861
Takeda M, Maeda H (1983) J Atmos Terr Phys 45:401
Takeuti T, et al (1978) J Geophys Res 83:2385
Tarpley JD (1970) Planet Space Sci 18:1091
Tatnall ARL, et al (1983) Space Sci Rev 35:139
Taylor WL (1960) J Res Nat Bur Stand 64D:349
Taylor WL (1969) J Atmos Terr Phys 31:983
Taylor WL, Sao K (1970) Radio Sci 5:1453
Thomas L (1982) In Rawer K (ed) Handbuch der Physik, 49(6). Springer, Berlin, Heidelberg, New York, p 7
Thorne RM (1975) Rev Geophys Space Phys 13:291
Uman MA (1969) Lightning. McGraw-Hill, New York
Uman MA, Krider EP (1982) IEEE Transact Electromagn Compat EMC-24:79
Uman MA, et al (1975) J Geophys Res 80:373
Uman MA, et al (1982) IEEE Transact Electromagn Compat EMC-24:410
Untiedt J (1967) J Geophys Res 72:5799
Untiedt J, et al (1978) J Geophys 45:41
Vasylinuas VM (1970) In McCormac BM (ed) Particles and Fields in the Magnetosphere. Reidel, Dordrecht, Holland, p 60
Vasyliunas VM (1975) Rev Geophys Space Phys 13:303
Volland H (1968) Die Ausbreitung langer Wellen. Vieweg, Braunschweig
Volland H (1976) J Geophys Res 81:1621
Volland H (1978) J Geophys Res 83:2695
Volland H (1982) In Volland H (ed) Handbook of Atmospherics, Vol I. CRC Press, Boca Raton, Fla, p 180
Volland H, Mayr HG (1977) Rev Geophys Space Phys 15:203

Volland H, et al (1983) J Geophys Res 88:1503
Vonnegut B (1953) Bull Am Meteorol Soc 34:378
Vonnegut B (1982) In Volland H (ed) Handbook of Atmospherics, Vol I. CRC Press, Boca
   Raton, Fla, p 1
Vonnegut B, et al (1966) J Atmos Sci 23:764
Wagner CU (1963) J Atmos Terr Phys 25:529
Wagner CU, et al (1980) Space Sci Rev 26:391
Wahlin L (1977) In Dolezalek H, Reiter R (eds) Electrical Processes in Atmospheres. Steinkopff,
   Darmstadt, p 384
Wait JR (1970) Electromagnetic Waves in Stratified Media. Pergamon Press, Oxford
Weidman CD, Krider EP (1978) J Geophys Res 83:6239
Whitten RC, Poppoff IG (1971) Fundamentals of Aeronomy. Wiley & Sons, New York
Wilcox JM, Scherrer PH (1972) J Geophys Res 77:5385
Williams DJ (1980) In Akasofu SI (ed) Dynamics of the Magnetosphere. Reidel, Dordrecht,
   Holland, p 407
Williams DJ (1983) Space Sci Rev 34:223
Willim DK (1974) In Holtet JA (ed) ELF-VLF Radio Wave Propagation. Reidel, Dordrecht,
   Holland, p 251
Winn WP, Byerley LG (1976) O J R Meteor Soc 101:979
Wolf RA, et al (1982) J Geophys Res 82:5257
Woodman RF, et al (1977) J Geophys Res 82:5257
Yasuhara F, Akasofu SI (1977) J Geophys Res 82:1279
Zanetti LJ, et al (1983) J Geophys Res 88:4875
Zirker JB (1977a) Rev Geophys Space Phys 15:257
Zirker JB (ed) (1977b) Coronal Holes and High Speed Wind Streams. Col Assoc Univ Press,
   Boulder, Col

Subject Index

Physics and Chemistry in Space

Editors: **L. J. Lanzerotti,**
J. T. Wasson

Springer-Verlag
Berlin
Heidelberg
New York
Tokyo

Physics and Chemistry in Space

Editors: L. J. Lanzerotti, J. T. Wasson

Springer-Verlag
Berlin
Heidelberg
New York
Tokyo